An Introduction to
Semiconductor Electronics

An Introduction to Semiconductor Electronics

K. J. CLOSE, B.SC., PH.D., F.INST.P.

Principal Lecturer in Physics,
The Polytechnic of Central London

J. YARWOOD, M.SC., F.INST.P.

Professor of Physics Emeritus,
The Polytechnic of Central London

Second Edition

Heinemann Educational Books
London

Heinemann Educational Books Ltd
22 Bedford Square, London WC1B 3HH

LONDON   EDINBURGH   MELBOURNE   AUCKLAND
HONG KONG   SINGAPORE   KUALA LUMPUR   NEW DELHI
IBADAN   NAIROBI   JOHANNESBURG
PORTSMOUTH (NH)   KINGSTON   PORT OF SPAIN

ISBN 0 435 68082 X

First published entitled *An Introduction to Semiconductors* 1971
Second edition entitled *An Introduction to Semiconductor Electronics* 1982
Reprinted with corrections 1985

British Library Cataloguing in Publication Data

Close K.J.
   An introduction to semiconductor electronics.—
   2nd ed.
   1. Semiconductors
   I. Title   II. Yarwood, J.
   621.3815′28   TK7871.85

The cover picture illustrates the use of a cluster of probes monitoring a thin film hybrid circuit (*courtesy of Philips Electronic and Associated Industries Ltd*)

Set in 10/11 Monophoto Times by
Mid-County Press, London
Printed and bound in Great Britain by
Biddles Ltd, Guildford and King's Lynn

Preface to First Edition

The junction diode, the various types of transistor, and other semiconductor devices have largely replaced the thermionic vacuum tubes, and especially the small power tubes, in the majority of applications in electronics.

It is relatively easy to give a simple account of the radio valves such as the triode and the pentode together with their easier circuit applications which can be understood by the beginner. It is not so easy to do this for the semiconductor devices. Yet it is imperative, even at the level of elementary physics in school, that this should be done, because otherwise the student's appreciation of modern electronics will be hopelessly out of date.

The syllabuses in physics of the various 'O' Level, 'A' Level, Ordinary National Certificate, and City and Guilds examinations are therefore desirably beginning to include elementary aspects of semiconductor physics. This tendency will inevitably increase in the future.

It is considered important that the student should not simply accept categorical statements about these devices, but gain a reasonably fundamental understanding of their operation which is based as far as is practicable on other aspects of physics in the course. The study should therefore be based on fundamental principles not only because of the educational value of such a procedure but also because this is the only way to ensure that the student will be able to understand the transistor device of a decade hence, likely to be significantly different from the ubiquitous junction transistor of today. The course should also include relevant simple calculations, exercises and simple practical experiments, yet must avoid the complexities of sophisticated topics such as Fermi-Dirac statistics and the quantum mechanics of conduction processes.

It is with these aims in view that this book is presented. It has been written with a paramount idea in mind: that a student at 'A'-Level standard should be able to understand it. It is not expected, however, that all the material included should be dealt with in a school course; some of it overlaps into the first year of science or technology at college or university.

It is hoped that this text will assist beginners in appreciating the fascination and utility of these devices which have changed electronics so greatly and will undoubtedly continue to do so in the future.

1971   K.J.C.
                                                                           J.Y.

Preface to Second Edition

During the decade since the first edition of this book was published, it is undoubtedly the case that there has been a considerable and a desirable increase in the interest in teaching electronics based on semiconductors in schools and colleges. In the same period, the spending of time on the principles of semiconductor physics has diminished. This is due primarily to the advent of the integrated circuit. It is now not so necessary to understand the active device and to obtain a comprehensive knowledge of the way in which working semiconductor electronic circuits have developed and are used. The emphasis is now on microelectronics and particularly pulse circuits, logic circuits and computation methods generally. To try to reflect this change — and at the same time to take into account the desirability in education of understanding the principles of a subject — the title of this book has been altered and the contents have been very considerably revised.

1982   K.J.C.
                                                                           J.Y.

Contents

1 The structure of matter and the solid state

1.1 Introduction

The study of semiconductors forms a part of the wider subject known as solid-state physics. The concern is with the passage of electric current through solids as opposed to liquids and gases.

The fundamental electric charge is that of the electron, a particle of mass m_e approximately 9×10^{-31} kg and charge e equal to 1.6×10^{-19} coulomb. The number of electrons in one coulomb of electric charge is thus $1/(1.6 \times 10^{-19})$, equal to 6.3×10^{18} approximately. As the ampere, the unit of electric current, corresponds to the passage of one coulomb of charge per second, so one ampere represents a transfer of 6.3×10^{18} electrons per second.

In this text the electron will be treated as a particle, even though on many occasions in solid-state physics, the location and behaviour of electrons have to be inferred by regarding the moving electron as a wave packet, leading to the methods of wave mechanics originated by the French physicist Louis de Broglie in 1924.

Once the existence of the electron was established finally in 1897 by Sir J. J. Thomson, much attention was devoted to explain the electrical properties of materials by the motion of electrons. Of particular interest were the metals, known to be very good conductors and therefore assumed to contain many 'free' electrons available for conduction, where by 'free' is meant that these electrons are bound in the metal by such insignificant forces that they are set in motion even by the smallest applied electric fields. Insulators were assumed to have no or very few free electrons available within their structure, whilst semiconductors, with electrical conductivities intermediate between those of metals and insulators, were represented by some intermediate state of electron containment.

In the case of metals, each atom may contribute one, two or even three free electrons to serve as current carriers. If each atom contributes only one free electron, there will be as many as 10^{29} free electrons per cubic metre of the metal. These electrons are able to move through the metal on the application of an electric field, but their motion will be restricted because they collide with the atoms of the metal.

1

An understanding of electrical conduction in solids thus demands a study of the electron as a charge carrier, the part it plays in atomic structure, the nature of interatomic forces and the arrangements of atoms in materials.

1.2 The Three States of Matter

A concentration of the atoms of an element can be arranged to be in a solid, liquid or gaseous state simply by controlling the environment in which the element is situated. In a solid element, the atoms are held in position by very strong mutual attractive forces which are electrical in origin. Their positions relative to one another in the solid at a given temperature will be in equilibrium when the system is in its lowest energy state. Work must be done to distort this equilibrium arrangement. This work is needed to alter the spacing between atoms. Because the displaced atoms will tend to return to their equilibrium positions, the solid is elastic. For example, the forces required to stretch a metallic rod show that the attractive forces between its atoms are considerable. Again, if any attempt is made to compress a solid, an elastic behaviour indicates the very strong repulsive forces between atoms which are brought into play when the spacings between atoms at equilibrium are decreased by the compression.

The atoms of a solid vibrate about their equilibrium positions with an amplitude which increases with temperature. An increase in temperature at first results in thermal expansion but, at the melting point, the rigid structure breaks down, the latent heat of fusion representing the energy needed to convert the solid into a liquid.

In the liquid state, which is the most complex of the three states of matter to study, the atoms move in random fashion, yet are still subjected to mutual attractions which retain the constant volume of the sample. That the motion of the atoms is very rapid is demonstrated by the ease with which diffusion occurs in liquids; diffusion also takes place in solids but at an almost imperceptible rate because of the constraint imposed by neighbouring atoms held in the rigid structure. Further evidence of the mutual attractive forces between atoms in a liquid is provided by the existence of surface tension forces.

At any temperature, whether the element is in a solid, liquid or gaseous state, the energies of the atoms are distributed about some average energy, which increases with the temperature. At any instant of time, some atoms will have energies much greater and some much less than this average value. Any fast-moving atoms near the surface of a solid or liquid may escape into the space above; solids and liquids therefore exert a vapour pressure, though that of liquids is normally much larger than that of solids. On evaporation from a liquid, only the most energetic atoms can escape, so that the average energy of all those atoms remaining behind must fall. The liquid temperature will therefore fall, and hence a liquid cools on evaporation. Further increase of temperature of the element beyond its melting point will increase the rate of evaporation. At the boiling point, which increases with the external pressure, the liquid is converted into vapour; this process requires an amount of heat called the latent heat of vaporization of the liquid. The forces of attraction between the atoms of the vapour are now very small compared with those which prevailed in the liquid state, and the vapour will now fill any container in which it is placed. The atoms can now travel significant distances between collisions, and these distances increase as the pressure is reduced. The motions of the atoms and their behaviour in the gaseous state are more readily understood than in the solid or liquid state.

In this account, atoms of an element have been discussed: similar statements apply to the molecules of a compound. The metallic elements have monatomic molecules, but an element such as hydrogen, oxygen or nitrogen will normally exist in the form of diatomic molecules.

1.3 Crystalline Structure

Many common substances are crystalline. Familiar examples are common salt (NaCl) and sugar. Each particle is a crystal having a specific geometrical form. Among the large crystals occurring in nature are quartz, diamond and mica. Metals are crystalline. In general, they are made up of a very large number of tiny crystals so small that a microscope is needed to discern them, though large crystals of metals can be grown by special techniques. X-rays can be used to examine the structure both of large single crystals and even of microcrystals which are not easily identified, even though discernible, under a microscope.

In a crystal, the atoms are arranged in a very orderly geometrical pattern known as the *crystal lattice*. All pure crystals of the same element have the same lattice structure. Solids in which there is no such definite arrangement are said to be *amorphous*, an example being glass.

The separation between neighbouring planes of atoms often differs in different directions in the crystal; thus, a different value for the elastic modulus can be measured in each of these directions. A specimen which exhibits different properties in different directions is called *anisotropic*. However, whereas single crystals are highly anisotropic, a specimen made up of millions of tiny crystals orientated at random, i.e. a polycrystalline specimen, is, in general, isotropic, a familiar example being an ordinary piece of copper, aluminium or iron.

There are seven basic patterns of crystal of which only one, the cubic, is considered here. Many materials have a cubic crystal structure. Few have the simple cubic structure of Figure 1.1(a), But this is considered here, for simplicity.

The simple cubic crystal is formed from an enormous number in adjacent cubic elements each of which is of the form shown of Figure 1.1(a). At each corner of the elemental cube there is an atom and these atoms are arranged to be alternately positive ions and negative ions.

The dimensions of such an elemental cube are very small. For example, the side of the cube may be only 0.282 nm (1 nm = 10^{-9} m

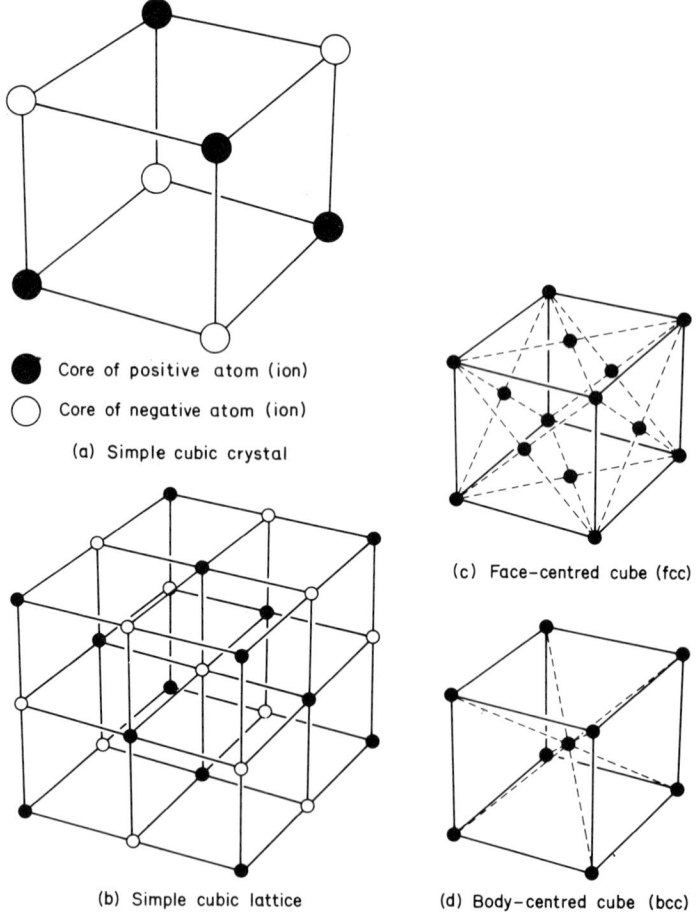

● Core of positive atom (ion)

○ Core of negative atom (ion)

(a) Simple cubic crystal

(c) Face-centred cube (fcc)

(b) Simple cubic lattice

(d) Body-centred cube (bcc)

Figure 1.1. Cubic crystal structures

= 10 Ångstrom unit). Even in a tiny microcrystal there are several million such elemental cubes, and these cubes are arranged in the crystal as shown in Figure 1.1(b), forming a crystal lattice in which it is noted that each atom (or rather ion) is at a point forming the corner of eight adjacent cubes.

The crystal lattice in this form is known as *simple cubic*. *Face-centred cubic* (fcc) and *body-centred cubic* (bcc) structures (Figure 1.1(c) and (d) respectively) are also frequently encountered. Many metals have crystals of one of these forms, for example, tungsten (bcc), molybdenum (bcc) and platinum (fcc).

Imperfections in a crystal influence greatly its physical and especially its electrical properties. A synthetic gemstone has only a fraction of the monetary value of a natural gemstone and yet it is a more perfect crystal. The attractive colour and lustre of a natural stone results from the presence of impurity atoms in the lattice.

If a foreign atom occupies a lattice site normally occupied by the host atom, it is called a *substitutional impurity*. On the other hand, if the foreign atom is lodged between the lattice planes of the host atoms it is said to be an *interstitial impurity*.

To change the electrical properties of a semiconducting crystal in a controlled fashion, foreign atoms of a selected element are introduced into substitutional sites in the host lattice. The techniques of introducing such foreign atoms into the pure crystal are described in Chapter 2.

1.4 The Motion of Free Electrons in an Electric Field

Because, by definition, the electric field strength at a point is the force on unit positive electric charge placed at any point, the force **F** (the bold print denotes a vector quantity) acting on a positive charge q in an electric field of strength **E** is given by

$$\mathbf{F} = q\mathbf{E} \qquad (1.1)$$

where the force acts in the direction of the field. The force is in newton if q is in coulomb and **E** is in volt per metre.

If V is the potential at this point, the field strength $\mathbf{E}_x$ is related to the potential gradient by

$$\mathbf{E}_x = \frac{-\mathrm{d}V}{\mathrm{d}x} \qquad (1.2)$$

where $\mathbf{E}_x$ is the component of the field strength in the x-direction. Therefore

$$\mathbf{F}_x = -q\frac{\mathrm{d}V}{\mathrm{d}x} \qquad (1.3)$$

For an electron with a negative charge of magnitude e the force is

$$F_x = e\frac{dV}{dx} \tag{1.4}$$

and is in the direction of V increasing. Under the action of this force, a free electron moves and gains energy. The work done by the force in moving it from point 1 at potential V_1 to point 2 at potential V_2 is seen from equation (1.4) to be

$$\int_1^2 F_x dx = \int_{V_1}^{V_2} e\, dV = eV_2 - eV_1 \tag{1.5}$$

This work done will give kinetic energy to the electron. The potential energy of the electron in position 1 is eV_1 and in position 2 is eV_2. If v_1 and v_2 are the speeds of the electron at the points 1 and 2 respectively,

$$eV_2 - eV_1 = \tfrac{1}{2}m_e v_2^2 - \tfrac{1}{2}m_e v_1^2 \tag{1.6}$$

If the electron starts from rest and moves through a potential difference of V, it acquires a speed v given by

$$eV = \tfrac{1}{2}m_e\, v^2 \tag{1.7}$$

This equation is only valid when the speed v of the electron is small compared with the speed of light. In the above it is assumed that the mass m_e is constant. In fact, in accordance with the Special Theory of Relativity, the mass of a particle increases with its speed but this increase is only of major importance as the speed approaches that of light in free space. The correction can usually be neglected if the electron is accelerated through less than 5 kV.

Example 1.4

Calculate the speed of electrons which are accelerated from rest through a potential difference of 5 kV.

(Ratio of charge to mass, e/m_e, for the electron $= 1.76 \times 10^{11}$ coulomb per kilogram.)

From equation (1.7)

$$v = \sqrt{\frac{2Ve}{m_e}}$$

Substituting $V = 5000$ and $e/m_e = 1.76 \times 10^{11}$ C kg^{-1},

$$v = \sqrt{(10\,000 \times 1.76 \times 10^{11})} \text{ metre per second}$$
$$= 4.2 \times 10^7 \text{ m s}^{-1}.$$

1.5 The Nuclear Model of the Atom: Ionization

The electrical and magnetic properties of materials are due primarily to the motions of electrons in the material. To understand the electrical properties of a solid it is consequently necessary to examine the atomic structure and also the arrangements of atoms within crystals.

The planetary or nuclear model of the atom, first described by Lord Rutherford in 1911, may be likened in some respects to our solar system. Around the positively charged nucleus, electrons rotate in certain orbits. These orbits are ellipses with the nucleus at one focus because the force of attraction between unlike charges (the positive nucleus and the negative electron) obeys the inverse square law.

In the neutral atom, the total number of electrons rotating in orbits around the nucleus is equal to the number of protons in the nucleus, which is Z, the atomic number. If an atom loses one or more of its orbital electrons, it is positively charged, and indeed becomes a positive ion. An atom which gains one or more electrons becomes a negative ion.

For ionization to occur, energy must be imparted to the atom to dislodge an electron from its outer orbital structure. This comes about by the incidence upon the atom of a sufficiently energetic particle. The chief ways of causing ionization are by the incidence of electrons and, less effectively, by irradiation with photons, the 'particles' of electromagnetic radiation which have an energy $h\nu$ where ν is the frequency of radiation and h is the Planck constant (6.6×10^{-34} joule second). Radiations from radioactive materials, in particular alpha-particles, beta-particles (high speed electrons) and, to a small extent, gamma rays, are also capable of causing ionization. Again, a material can be ionized by heating it to a sufficiently high temperature (e.g. gas in a flame).

The atoms of some solid semiconductors can be ionized by the incidence of visible light. These semiconductors are used to advantage in solid-state photoelectric cells.

Example 1.5(a)

Calculate the energy in electron-volt of a photon in monochromatic light of wavelength 550 nm. (The speed of light in free space $= 3 \times 10^8$ m s^{-1}, the Planck constant $h = 6.6 \times 10^{-34}$ Js.)

The energy E of the photon of frequency ν is given by

$$E = h\nu = hc/\lambda$$

where c is the speed of light in free space and λ is the wavelength corresponding to a frequency ν. Substituting the values given

$$E = \frac{6.6 \times 10^{-34} \times 3 \times 10^8}{550 \times 10^{-9}} = 3.6 \times 10^{-19} \text{ joule}$$

As $1 \text{ eV} = 1.6 \times 10^{-19}$ joule, therefore

$$E = \frac{3.6 \times 10^{-19}}{1.6 \times 10^{-19}} = 2.25 \text{ eV}$$

Example 1.5(b)

The radius of the electron orbit (assumed to be circular) in a hydrogen atom (atomic number $Z = 1$) is 5.3×10^{-2} nm. Calculate the speed with which the electron describes the orbit. (For the electron the mass is 9.1×10^{-31} kg and the charge is 1.6×10^{-19} C.)

To retain the electron of mass m_e in a circular orbit of radius r with a peripheral speed v, a force of magnitude $m_e v^2 / r$ is required to be directed towards the central nucleus. This is provided by the force of attraction between unlike charges. As the nuclear charge is Ze, where e is positive and numerically equal to the electron charge, therefore

$$\frac{Ze \cdot e}{4\pi\varepsilon_0 r^2} = \frac{m_e v^2}{r}$$

where ε_0 is the permittivity of free space. For the hydrogen atom $Z = 1$, and therefore

$$v = \sqrt{\frac{e^2}{4\pi\varepsilon_0 m_e r}}$$

Substituting the values given together with $\varepsilon_0 = 8.854 \times 10^{-12}$ farad per metre,

$$v = \sqrt{\frac{(1.6 \times 10^{-19})^2}{4\pi \times 8.854 \times 10^{-12} \times 9.1 \times 10^{-31} \times 5.3 \times 10^{-11}}} \text{ m s}^{-1}$$

$$= 2.2 \times 10^6 \text{ m s}^{-1}.$$

Example 1.5(c)

Calculate the ionization potential of atomic hydrogen.

Ionization takes place when the electron is removed from the normal ground state (the state when it is as close to the nucleus as it can be) to outside the influence of the nucleus of the hydrogen atom; such detachment corresponds to removing the electron from a distance r (the radius of the circular orbit in the ground state) to infinity.

Assuming a point at an infinite distance from the nucleus to be at zero potential energy, an electron moving from infinity to a distance r from the nucleus of positive charge Ze requires an amount of work to be done, so acquires a potential energy E_p given by

$$E_p = \int_{\infty}^{r} \frac{Ze^2}{4\pi\varepsilon_0 r^2} \, dr$$

because work = force × distance and the force on an electron at a distance r is $Ze^2/4\pi\varepsilon_0 r^2$. Hence

$$E_p = -\left[\frac{Ze^2}{4\pi\varepsilon_0 r}\right]_\infty^r = \frac{-Ze^2}{4\pi\varepsilon_0 r}$$

Note that this potential energy E_p is negative. This is because the potential energy of an electron at an infinite distance from the nucleus is taken to be zero, and work has to be done *on* the electron to *remove* it from the attractive effect of the positive nucleus.

However, the electron of mass m_e also possesses kinetic energy E_K of $\frac{1}{2}m_e v^2$ where v is its peripheral speed which, at the distance r, is given from Example 1.5(b) to be

$$E_K = \tfrac{1}{2}m_e v^2 = Ze^2/8\pi\varepsilon_0 r$$

The total energy of the electron when in the circular orbit of radius r is therefore

$$E = E_K + E_p$$

$$= \frac{Ze^2}{8\pi\varepsilon_0 r} - \frac{Ze^2}{4\pi\varepsilon_0 r} = \frac{-Ze^2}{8\pi\varepsilon_0 r}$$

The negative sign arises simple from the fact that the reference point of zero potential energy is conventionally taken to be at infinity.

For hydrogen, $Z = 1$, therefore

$$E = e^2/8\pi\varepsilon_0 r$$

$$= (1.6 \times 10^{-19})^2/(8\pi \times 8.854 \times 10^{-12} \times 5.3 \times 10^{-11})$$

on substituting $e = 1.6 \times 10^{-19}$ C, $\varepsilon_0 = 8.854 \times 10^{-12}$ F m^{-1} and $r = 5.3 \times 10^{-11}$ m (see Example 1.5(b)). Hence

$$E = 2.2 \times 10^{-18} \text{ J} = \frac{2.2 \times 10^{-18}}{1.6 \times 10^{-19}} \text{ eV} = 13.7 \text{ eV}$$

because 1 eV $= 1.6 \times 10^{-19}$ joule. The ionization potential, defined as the minimum potential difference through which an electron must fall to acquire sufficient energy to ionize the material, is therefore 13.7 eV for hydrogen.

1.6 Motion of Electrons in a Uniform Magnetic Field

Consider a beam of electrons of uniform speed v passing through a region of uniform magnetic flux density directed perpendicularly to the path of the beam (Figure 1.2). If there are n electrons each of charge e in a metre length of the beam each travelling with a speed v in a vacuum, the rate of passage of charge at any point in the beam is nev, which is the current I. Therefore, the magnitude of the force δF due to the magnetic flux of density B on the electrons with a path length δs is given by

$$\delta F = BI . \delta s = Bnev . \delta s \tag{1.8}$$

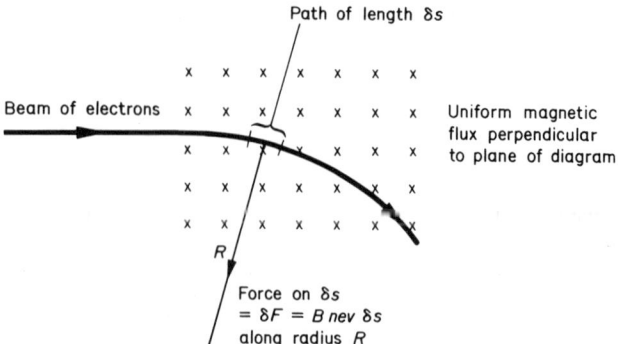

Figure 1.2. Passage of a beam of electrons of uniform speed through perpendicular uniform magnetic flux

So the force on each electron, of which there are $n\delta s$ in the length δs, is Bev. In accordance with Fleming's left-hand rule, this force is directed normally to the directions of both B and δs. If the beam makes an angle θ with the direction of the magnetic flux, the force on each electron becomes

$$F = Bev.\sin\theta \qquad (1.9)$$

Three important considerations emerge from equation (1.9).

(a) When $v = 0$, $F = 0$; there is no force on a stationary electron in magnetic flux. In an electric field the electron would, on the other hand, be accelerated due to the force acting (section 1.4).

(b) The force is a maximum when $\theta = 90°$, i.e. the electron moves at right angles to the direction of the magnetic flux, which is the direction of B. If $\theta = 0°$, $F = 0$ so there is no force acting on an electron moving along the magnetic flux lines.

(c) As the force is perpendicular to the velocity of the electron it can change the direction of motion but cannot affect its energy. Thus, an electron cannot gain energy from static magnetic flux.

For the special case of $\theta = 90°$ the force is always normal to the electron path which will cause the electron to move in a circular orbit at constant speed. Suppose this circle or arc of a circle in the magnetic field has a radius R (Figure 1.2). An electron travelling with constant speed v in a circular path of radius R has an acceleration towards the centre of the circle of v^2/R, the corresponding force being the mass times the acceleration, i.e. $m_e v^2/R$. This equals the force due to the motion of the charged particle in the magnetic flux, which is Bev.

$$Bev = m_e v^2/R \qquad (1.10)$$

or $$v = BeR/m_e$$

The time period T for one revolution of the electron is $2\pi R/v$. Hence

$$T = 2\pi m_e v/Bev = 2\pi m_e/Be \qquad (1.11)$$

It is apparent that these equations (1.10) and (1.11) apply also to other charged particles than the electron where, in general, the electric charge is q and the mass is m.

1.7 The Arrangement of Electrons in Atoms: The Periodic Table

The elements occurring in nature range in atomic numbers from hydrogen $(Z = 1)$ to uranium $(Z = 92)$. Values for some of the intermediate elements of importance in this book are carbon $(Z = 6)$, sodium $(Z = 11)$, silicon $(Z = 14)$, copper $(Z = 29)$ and germanium $(Z = 32)$.

An element such as copper is a very good conductor of electricity and has a positive temperature coefficient of resistance. This conductivity is due to the motion of electrons (within the copper) which move in an applied electric field. The atoms are within a face-centred cubic crystal structure. If the mechanism of this conduction is to be explained more intimately, the arrangement of the electrons in the electronic structure of the copper atom needs to be described more fully. Knowledge about this arrangement is also necessary to explain, for example, the semiconducting properties of silicon and why a compound such as aluminium oxide is an excellent insulator.

The experimental study of the line spectra of elements is the chief source of circumstantial evidence about atomic structure. All the evidence from spectroscopic and allied studies leads to the conclusion that each and every kind of atom has a specific structure.

For example, copper has an atomic number of 29 and a mass number (the nearest integer to its atomic mass) of 63 in its most abundant form. The nucleus of the copper atom thus has a positive charge of $29e$, where e is numerically equal to the electronic charge, due to the 29 protons in its nucleus. Within this nucleus there are also $(63 - 29) = 34$ neutrons. Around the nucleus in the neutral atom circulate 29 electrons.

The proton and the neutron have nearly equal masses and each is nearly 2000 times the mass of the electron. The massive very tiny central nucleus will therefore not be subject to any significant disturbance in the process of electric conduction or the emission of spectral radiation, ordinary chemical reactions and so on. It is to the electrons outside the nucleus that we must look to explain such phenomena.

In the copper atom, for example, each and every one of the 29 electrons circulating the nucleus has a most probable location with respect to the nucleus. Is there any manner in which these locations can be specified? The answer is a qualified yes. A vast amount of experimental evidence about the behaviour of the atoms of the elements can be explained on the basis that each electron in the atom is in a specific state (fully designated in the Pauli exclusion principle by a set of four quantum numbers, which will not be discussed here) and that no two electrons in a given atom can be in the same state. These concepts lead to the valuable picture of the atom in which electrons are in specific groups called main *shells* and *sub-shells* where the sub-shell is a part of a main shell. The main shells, numbered $n = 1$, 2, 3, 4, 5 proceeding 'outwards' from the nucleus are called successively the K, L, M, N, and O shells respectively. By 'outwards' is meant in a simple model of the atom, further from the nucleus. More specifically, the shells of lower number n (called the principal quantum number) are the more closely bound to the nucleus and electrons in them have to receive more energy to dislodge them from the atomic structure. Thus, in spectroscopy, photons of energies in the x-ray region have to be used to dislodge electrons from the K and L shells of copper whereas those in the outermost shells are 'free' to take part in electrical conduction and in the formation of copper compounds such as the oxide.

For a given main shell number n, it is found that the maximum number of electrons it can contain is $2n^2$. Furthermore, when a given shell contains its full complement of electrons it is said to be 'closed' or 'filled'. A filled shell forms a particularly stable arrangement of electrons, not easily disturbed by external influences. For example, for $n = 2$, $2n^2 = 8$. If this second shell *does* contain 8 electrons it is extra-stable and electrons are not so readily extracted from it to take part in physical and chemical phenomena as if it were to contain less than 8 electrons. Thus we get Table 1.1.

The familiar element sodium ($Z = 11$) will have 11 electrons outside the nucleus in its atom. The innermost shells will be filled because the electrons will occupy allowed states as closely bound to the nucleus as possible. For sodium, the K shell and the L shell will be filled with 2

Main shell number n	1	2	3	4	5
Letter designation	K	L	M	N	O
Number of electrons in filled shell	2	8	18	32	50

Table 1.1.

and 8 electrons respectively, leaving the eleventh electron in an unfilled M shell. This lonely outermost electron is at the mercy of external forces applied to sodium. It is the electron responsible for the conduction of sodium, for the visible emission spectrum (the familiar lines at 589 and 589.6 nm in the sodium spectrum) and is the electron which is readily transferred to another element in the formation of a chemical compound (e.g. sodium chloride) so it is the *valence electron*.

The sub-shells are designated by a second quantum number l which, like n, is also an integer*. Within a main shell n, the possible values of l are 0, 1, 2, 3 etc. up to $(n-1)$. Again, designation by letter notation is used. The lower-case letters employed are s, p, d, f, g corresponding to l values of 0, 1, 2, 3, 4 respectively. These letters are chosen for somewhat unsatisfactory historical reasons arising from the early days of observations on the emission line spectra of elements: they are respectively the initial letters of the descriptive terms *sharp, principal, diffuse* and *fundamental*; beyond f, further letters are in alphabetical order.

A sub-shell characterized by an integer l is filled when it contains $2(2l+1)$ electrons. For example, the letter d corresponds to $l=2$; a d sub-shell is filled when it contains $2(2 \times 2 + 1) = 10$ electrons. Filled sub-shells are also stable configurations within the main shells though their stabilities are not so marked as those of main shells.

On this basis Table 1.1 can be extended to give Table 1.2. This table which gives the maximum numbers of electrons that can exist in the shells and sub-shells of atoms (based on the Pauli exclusion principle which is at the heart of the considerations which lead to the numbers quoted), is of the utmost value because it forms the basis of the explanation of the electronic structure of the elements in the periodic table.

Main shell (letter)	K	L		M			N				O				
Main shell number (n)	1	2		3			4				5				
Number of electrons in filled shell ($2n^2$)	2	8		18			32				50				
Sub-shells within main shell (l)	0	0	1	0	1	2	0	1	2	3	0	1	2	3	4
Letter designation of l	s	s	p	s	p	d	s	p	d	f	s	p	d	f	g
Number of electrons in filled sub-shell $[2(2l+1)]$	2	2	6	2	6	10	2	6	10	14	2	6	10	14	18

Table 1.2.

* This assumption has to be modified in more advanced theory.

In Table 1.3 is given such detail for the first 37 elements from hydrogen ($Z = 1$) to rubidium ($Z = 37$). The horizontal lines on this table mark the closure of a shell or sub-shell arrangement.

In relation to this Table 1.3 note the closed shell or sub-shell structure corresponding to the inert gases helium ($Z = 2$), neon ($Z = 10$), argon ($Z = 18$) and krypton ($Z = 36$).

Some of the elements to which reference will be made later in this text are sodium, copper, aluminium, carbon, silicon, germanium, chlorine and oxygen. Consider these elements in relation to Table 1.3:

Sodium ($Z = 11$): a single electron in the 3s state (main shell $n = 3$; sub-shell $l = 0$, corresponding to s) outside a filled shell structure like that of neon. This single electron is the valence electron; it is also the electron which is 'free' to take part in electrical conduction. When it leaves the atom, the remaining structure is a positive ion; sodium is an electropositive element of valence unity, i.e. is monovalent.

Copper ($Z = 29$): a single electron in the 4s state outside filled shells K (2 electrons), L (8 electrons) and M (18 electrons). This 'free' electron, shielded from the nucleus by 28 electrons, is the main conduction electron (Figure 1.3).

Aluminium ($Z = 13$): three electrons outside a filled shell structure like that in neon. Two of these electrons are in the 3s state and one in 3p. A good conductor which exhibits trivalent activity when all these three electrons enter in chemical combination, as in aluminium oxide, Al_2O_3.

Carbon ($Z = 6$): four electrons in the L shell outside the filled K shell. These four are half the full complement possible. Carbon is intermediate in electrical and chemical behaviour between electropositive metallic lithium and the electronegative gas fluorine ($Z = 9$) which requires only one electron to complete the outermost shell. Carbon has semiconducting properties and, like silicon and germanium, a negative temperature coefficient of resistance. Its maximum valence is four, i.e. quadravalent.

Silicon ($Z = 14$) and *germanium* ($Z = 32$): similar to carbon because the first has four electrons outside closed K and L shells and the second has four electrons outside filled K, L and M shells. These elements, especially silicon, are of prime importance in the semiconductor diode and the transistor.

Chlorine ($Z = 17$) and *fluorine* ($Z = 9$): both need only one electron to fill their shell structures. They are elements which avidly acquire electrons to become negative ions both in gases and in crystalline solids. Both are monovalent electronegative elements. Both combine

Element	Atomic no. Z	K 1s	L 2s	2p	M 3s	3p	3d	N 4s	4p	4d	4f	O 5s	5p	5d	5g
H	1	1													
He	2	2													
Li	3	2	1												
Be	4	2	2												
B	5	2	2	1											
C	6	2	2	2											
N	7	2	2	3											
O	8	2	2	4											
F	9	2	2	5											
Ne	10	2	2	6											
Na	11	Core	of 10		1										
Mg	12	Core	of 10		2										
Al	13	of 10			2	1									
Si	14	electrons			2	2									
P	15	as in			2	3									
S	16	neon			2	4									
Cl	17				2	5									
A	18				2	6									
K	19	Core	of 18					1							
Ca	20	Core	of 18					2							
Sc	21	Core	of 18				1	2							
Ti	22	Core	of 18				2	2							
V	23	of 18					3	2							
Cr	24	electrons					5	1							
Mn	25	as in					5	2							
Fe	26	argon					6	2							
Co	27						7	2							
Ni	28						8	2							
Cu	29						10	1							
Zn	30						10	2							
Ga	31						10	2	1						
Ge	32						10	2	2						
As	33						10	2	3						
Se	34						10	2	4						
Br	35						10	2	5						
Kr	36						10	2	6						
Rb	37	Core of 36 electrons as in krypton										1			

Table 1.3. Arrangements of electrons in the main shells and sub-shells of atoms

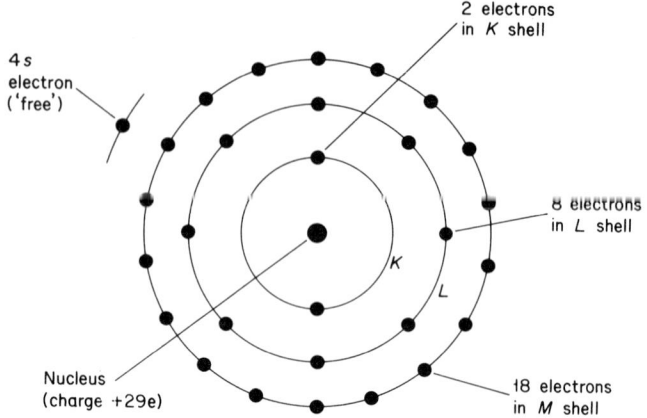

Figure 1.3. Electrons in shells in the copper atom (Schematic diagram)

very readily with monovalent electropositive alkali metals such as lithium, sodium and potassium.

Oxygen $(Z = 8)$: requires two more electrons to complete its unfilled L shell. It is usually a divalent electronegative element, tending to form negative ions in gases and solids.

1.8 The Conduction of Sodium

Consider the conduction of the metal sodium. As shown in Section 1.7 the sodium atom has a single electron in the M shell outside 10 electrons in closed K and L shells. These ten electrons provide an electrostatic screen between this outermost single electron and the positive nucleus with its charge of $11e$. This single electron in the unfilled M shell in an atom amongst myriads of atoms in a piece of sodium has no more definite affinity to any one atom than it does to the immediately neighbouring atoms. It is said to be 'free'. When an electric field is applied to sodium any free electrons therefore move readily along the electric field lines. In this respect they move in the space between the atoms in the crystal lattice of sodium metal (actually a body-centred cubical lattice) rather like molecules move in the space between molecules in a gas. The analogy can be carried further in that the restriction to the motion of the free electrons in sodium (and other metallic conductors) is brought about by collisions of these electrons with sodium atoms in the crystal lattice, comparing with the restriction on molecular motion in gases due to collisions between molecules.

1.9 Some Electrical Properties Characteristic of Metals

(*a*) Metals have resistivities with the range 10^{-8} to 10^{-5} Ω m. Silver, one of the best conductors, has a resistivity of 1.5×10^{-8} Ω m at room temperature whereas that of manganese is 2.6×10^{-6} Ω m.

(*b*) Metals have a very high electron density which does not change with temperature; thus the number of current carriers (free electrons) available is constant.

(*c*) Metals have a small positive temperature coefficient of resistivity of approximately $0.3\% \ K^{-1}$. A rise in temperature increases the amplitude of vibration of the lattice atoms which consequently retard more the drift motion of the free electrons in an applied electric field. The resistance of a metal therefore increases with temperature.

(*d*) The resistivities at room temperature of any two polycrystalline samples of the same metal are virtually identical even though the polycrystalline structures differ.

1.10 Semiconductors

For about 100 years, solids have been known with resistivities in the range 10^{-3} to 10^{4} Ω m. It is within this group that semiconductors exist. However, while some impure or heavily doped semiconductors can have electrical properties quite similar to metals, those in a very pure state may be good insulators, especially if maintained at low temperature.

Much of the early work on semiconductors carried out over many years prior to the discovery of the transistor produced very confusing results. The measured electrical properties of two apparently identical specimens were often quite different. At the time the influence of impurity atoms, the surface state of the sample and structural imperfections were not appreciated. Thus structural imperfections, in particular the grain boundaries between the tiny crystals of a polycrystalline specimen, interfere with the electron motion and so radically alter the electrical behaviour. It was not until the specimens were cut from large nearly perfect crystals that consistent and reproducible experimental results were obtained.

Some semiconductors of importance in electronics are listed below. Although some metallic salts, particularly the alkali metal halides (e.g. NaCl) show moderate ionic conductivity at high temperatures — which classifies them as semiconductors — they are at present of little interest in electronics so will not be discussed.

Elements   Silicon (Si), germanium (Ge), tellurium (Te), selenium (Se).

Compound semiconductors These are also called 'inter-metallic compounds' or '3–5' compounds. They are compounds of metals of which the component metals come from the third and fifth groups of the periodic table. Two much used examples are gallium arsenide (GaAs) and indium antimonide (InSb). Thus gallium has an atomic number Z of 31. It therefore has three electrons outside 28 electrons in the filled K, L and M shells which contain 2, 8 and 18 electrons respectively (Table 1.3). As a trivalent element it is group III of the periodic table. Arsenic has an atomic number Z of 33; it therefore contains 5 electrons outside the 28 electrons in the filled K, L and M shells (Table 1.3) and so is in the group V of the periodic table.

For reference purposes it should be noted that group III is divided into group IIIA containing the elements boron, aluminium, gallium, indium and titanium and group IIIB comprising scandium, yttrium, lanthanum and actinium; also group V consists of group VA (nitrogen, phosphorus, arsenic, antimony and bismuth) and group VB (vanadium, niobium and tantalum).

Note also indium in group III and antimony in group V. It is also significant that the three valence electrons of a metal in group III combine with the five electrons in a metal of group V to provide the eight electrons of a relatively stable shell structure.

Other inorganic compounds In particular we have cadmium sulphide (CdS), zinc sulphide (ZnS) and zinc oxide. Cadmium ($Z = 48$) and zinc ($Z = 30$) are both divalent metals in group II of the periodic table (e.g. zinc contains two electrons in the N shell outside the twenty-eight in the filled K, L and M shells, see Table 1.3) whereas sulphur ($Z = 16$) and oxygen ($Z = 8$) are in group VI (e.g. sulphur contains six electrons outside filled K and L shells, see Table 1.3). These semiconducting inorganic compounds are hence known as '2–6' compounds.

Apart from a conductivity within a certain range, semiconductors usually show electrical behaviour summarized as follows:

(i) A large negative temperature coefficient of resistance of approximately $6\% \ K^{-1}$.

(ii) Sensitivity to light: incident radiation produces a photovoltaic or photoconductive effect.

(iii) The electrical properties change markedly when the specimen is placed in a magnetic field. The current carriers are strongly influenced so that magnetoresistance (resistance change when placed in a magnetic field) and the Hall effect (section 1.18) are much larger than for metals.

(iv) Impurity atoms have a very marked effect. For example, a pentavalent impurity (group V element) added to the extent of only 1 part in 10^8 will increase the conductivity of germanium by 10 times at 300 K.

(v) Properties are considerably affected by the surface state of the sample. Molecules of water on the surface of a crystal specimen alter profoundly its electrical behaviour. Also when a contact is made between a metal and a semiconductor, it may be ohmic (i.e. obey Ohm's law) or have rectifying properties depending on the state of the interface at the contact. An ohmic contact will have linear characteristics, so that the current through the junction will be a linear function of the voltage across it. Ohmic contacts are often made by doping heavily the semiconductor where the metal is to be attached.

The electron gas model applied with modest success to metals is of little value in explaining the electrical behaviour of semiconductors. One of the triumphs of the quantum theory of solids in the 1930s was the explanation of the effects of current carriers in semiconductors. This work led to the energy band theory of solids which not only explained the temperature and impurity sensitivity of semiconductors but also enabled materials to be broadly classified into conductors, semiconductors and insulators in terms of their electronic or extra-nuclear atomic structure.

1.11 Pure Semiconducting Elements

The two important elements are silicon and to some extent germanium. These are alike in that silicon has 4 valence electrons outside filled K and L shells (2 in K and 8 in L) whereas germanium has 4 valence electrons outside filled K, L and M shells (18 in M). They are both quadravalent elements and are in group IV (strictly IVA) of the periodic table in which also occurs carbon ($Z = 6$) which has four valence electrons outside the two electrons in the filled K shell.

A crystalline form of carbon is diamond. Silicon and germanium both crystallize with a diamond-like structure. In the crystal lattice of silicon (represented in two dimensions in Figure 1.4) each of the four valence electrons of a particular atom is shared by one of the four nearest neighbouring atoms. With all four of the valence electrons used in the covalent bonding (as it is called) there are no free electrons available. At low temperature (near 0 K) silicon is therefore an insulator.

If the temperature of a silicon crystal is raised, the electrons gain energy from the lattice vibrations which increase with temperature. Some of the valence electrons may gain sufficient energy to break free from the covalent bonds and become available as current carriers.

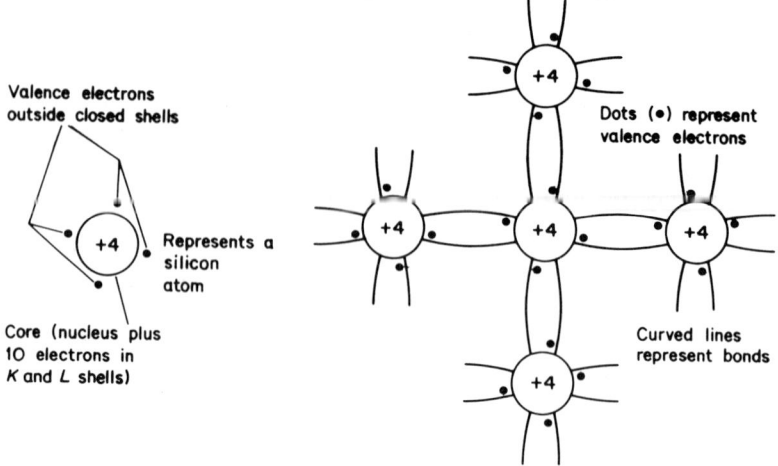

Figure 1.4. Atoms of silicon in the crystal lattice showing the covalent bonds between neighbouring atoms. (A two-dimensional schematic diagram of what is actually a three-dimensional array)

An electron so released from such bondage must leave behind a vacancy in the lattice. Such a vacancy is called a *hole*. This hole therefore exists within the valence electron structure of the atom. It cannot exist for any significant time because the positive nucleus of the atom concerned will attract an electron (able to move in an applied electric field) from elsewhere in the neighbouring atoms so the hole becomes filled, the normal valence structure of the atom becoming completed thereby. This must transfer the vacancy (hole) to the neighbouring atom which, in turn, passes its electron deficiency to the next neighbour, and so on. The hole is therefore mobile in an applied electric field, but hole motion will take place in the *opposite direction to electron motion*. The hole therefore has the property of a positive charge, and is known as a *positive hole*. Electron and positive hole motion in an applied electric field are illustrated schematically by Figure 1.5. It should be noted that the concept of positive hole conduction is a convenience; the reality is still the motion of electrons.

The semiconducting crystalline elements silicon and germanium in the pure state therefore exhibit electrical conductivity at elevated temperatures because of the generation within their lattices of *electron-hole pairs* where the electron is freed from covalent bonding by temperature rise and necessarily leaves behind it the hole or vacancy. The conductivity resulting from the generation of electron-hole pairs in pure silicon or germanium, free from lattice imperfections, is known as *intrinsic conductivity*. It is distinguished

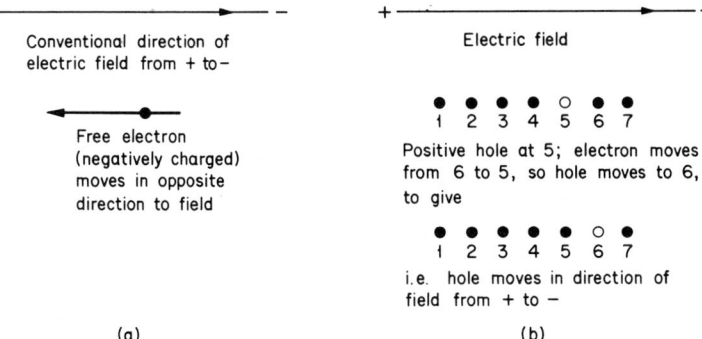

Figure 1.5. (a) Electron motion in an applied electric field and (b) positive hole motion in a semiconductor in an applied electric field

from *extrinsic conductivity* which results (section 1.12) from impurities in the crystal.

Intrinsic conductivity may be attributed to ionization produced thermally in the solid silicon or germanium. At any instant electrons and holes are being produced and are also combining with one another. At any given temperature, a state of dynamic equilibrium exists between the ionization and recombination processes.

1.12 Extrinsic or Impurity Conductivity

The conductivity of silicon can be increased, while still preserving the crystal lattice configuration, by the addition of certain impurities. Atoms are chosen which can fit into the lattice structure without inducing undue strain. The most useful impurities to introduce in a controlled fashion are selected elements with a valence of 5 (pentavalent) or 3 (trivalent), that is either one more or one less than the quadravalent silicon. The addition of such impurities is known as *doping*.

Pentavalent (i.e. group V) elements suitable for doping the silicon crystal are arsenic (As), phosphorus (P) or antimony (Sb). Reference to Table 1.3 shows, for example, that arsenic $(Z = 33)$ has 5 valence electrons outside the 28 electrons in the filled K, L and M shells which, together, contain $2 + 8 + 18 = 28$ electrons.

The effect of incorporating pentavalent phosphorus within the silicon lattice is shown in Figure 1.6(a). Four of the five valence electrons of phosphorus will take part in the covalent bonds. The fifth valence electron is bound only very weakly to its parent atom. Only a small amount of energy is required to free this fifth electron, indeed only about 0.05 eV as compared with the 1.1 eV required in pure silicon.

Before the temperature is high enough for appreciable intrinsic

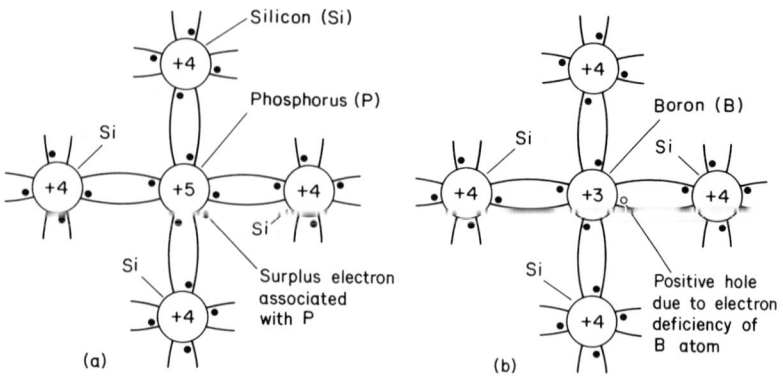

Figure 1.6. (a) One atom of silicon in the crystal lattice is replaced by pentavalent phosphorus leaving a surplus free electron and (b) one atom of silicon is replaced by trivalent boron leaving a surplus positive hole

conductivity of the silicon itself to occur, all the impurity atoms will have contributed a free electron and become ionized. The pentavalent phosphorus donates conduction electrons to the silicon so is called a *donor impurity*. The impure or doped silicon is called *n-type* because negative charges (i.e. electrons) are responsible for the increased conductivity consequent on pentavalent impurity additions.

On the other hand, if the added impurity is trivalent, for example, aluminium, boron or indium, the modified lattice structure is as shown in Figure 1.6(b). Aluminium ($Z = 13$), for example, is trivalent in group III of the periodic table because it has three valence electrons outside a closed shell structure of 10 electrons comprising 2 in the filled K shell and 8 in the filled L shell.

As an example, the widely used trivalent element boron ($Z = 5$) has one too few valence electrons to provide the four necessary for the covalent bonds in quadravalent silicon. There is consequently a vacancy or positive hole left in the lattice. This hole readily captures a thermally energized electron from a neighbouring atom and again this transfer requires less energy than that required for an electron to break free from the silicon bond.

Enhanced conduction is now primarily due to positive holes and the doped silicon is called *p-type*. The added impurity is known as an *acceptor* because it accepts an electron to fill the vacancy and complete the bonding.

A donor atom becomes a positive ion when it donates an electron whereas an acceptor atom becomes a negative ion when it gains an electron. Nevertheless, the total charge on the crystal is always zero because the ionization merely results from the production of electron-hole pairs. Ions formed in a doped crystal are locked in the lattice, otherwise they too would contribute to the conductivity. Ionization can be produced by the action of light and this process is used to advantage in the photodiode and the phototransistor (section 5.1). Transistors are mounted inside the light-tight capsules to exclude this interference.

Purity in semiconductors is a relative term. A sample of silicon which has been subjected to zone refining techniques (Chapter 2) might have an impurity concentration of 1 part in 10^{10}, corresponding to 10^{19} impurity atoms per cubic metre. This is termed a 'near-intrinsic specimen', although it will exhibit very low n- or p-type conductivity depending upon whether the residual impurity is pentavalent or trivalent. A doped n-type material will have electrons as *majority carriers*. Because some small quantity of trivalent impurity is inevitably present, some hole conduction must occur. These positive holes are called *minority carriers*.

It is possible to convert n-type to p-type or *vice versa* simply by adding sufficient impurity to outnumber the majority carriers, which will then be in the minority. In manufacturing processes, the material is refined to a near-intrinsic state and the impurity added to achieve a conductivity of the correct type and magnitude.

1.13 Carrier Mobility and Current Density

The motion of current carriers (electrons and positive holes) can occur as the result of

(a) drift in an applied field;
(b) diffusion from a region of high concentration to one of lower concentration.

Only the first of these processes is examined at present. At any given temperature the carriers move in random directions with a distribution of speeds and make collisions with the vibrating lattice ions. When an electric field is applied to the semiconductor, positive holes acquire a drift component velocity in the direction of the field whereas electrons drift in the opposite direction (this is using the convention that the direction of an electric field is that in which a positive charge would move or tend to move). Superimposed on the random motion, therefore, is a velocity component produced by the applied field.

If n_p is the density of positive holes (number of positive holes per cubic metre) and v_p metre per second is their drift velocity, the current density J_p, defined as the current through unit cross-sectional area perpendicular to v_p, due to positive holes is given by

$$J_p = n_p e v_p \text{ ampere per square metre} \tag{1.12}$$

where e is the electronic charge in coulomb, Similarly, the current density J_n resulting from electron motion in the direction of the applied field is given by

$$J_n = n_n e v_n \text{ A m}^{-2} \tag{1.13}$$

where the subscript n denotes electron. The mobility k of a current carrier is defined as the velocity acquired in unit electric field. Therefore

$$k_n = v_n/E \text{ and } k_p = v_p/E \tag{1.14}$$

the unit of mobility k being (metre per second)/(volt per metre), i.e. m s^{-1}/V m^{-1} or m^2 V^{-1} s^{-1}.

The total current density J is simply the sum of J_n and J_p. Hence, from equations (1.12), (1.13) and (1.14),

$$J = eE(n_n k_n + n_p k_p) \tag{1.15}$$

The conductivity σ is the current density in unit electric field, so

$$\sigma = e(n_n k_n + n_p k_p) \tag{1.16}$$

For a near intrinsic specimen, equal numbers of electrons and holes exist at any instant of time so $n_n = n_p = n_i$, where n_i is the density of current carriers in the intrinsic case. Equation (1.16) can then be written

$$\sigma = n_i e(k_n + k_p) \tag{1.17}$$

Example 1.13(a)

In a near intrinsic specimen of germanium at 300 K the electron density is 2 $\times 10^{19}$ m^{-3}. If the mobilities of the electrons and positive holes are respectively 0.39 and 0.19 m^2 V^{-1} s^{-1}, calculate the resistivity of germanium.

Using equation (1.17),

$$\sigma = 2 \times 10^{19} \times 1.6 \times 10^{-19}(0.39 + 0.19)$$

on putting the electronic charge $e = 1.6 \times 10^{-19}$ coulomb. Hence

$$\sigma = 1.86 \text{ S m}^{-1}$$

(S = siemen, the reciprocal of ohm.)

Resistivity $\rho = 1/\sigma = 1/1.86 = 0.54$ Ω m.

Example 1.13(b)

Calculate the mobility at 0°C of electrons in silver given that there are 10^{29} free electrons per cubic metre in silver and its resistivity at 273 K (0°C) is 1.5 $\times 10^{-8}$ Ω m.

The current density J in a silver specimen is given by

$$J = nev$$

where n is the current carrier density, and only electrons act as current carriers, e is the electronic charge and v is the drift velocity of the electrons in an electric field E. Therefore

$$\sigma = \frac{J}{E} = \frac{1}{\rho} = nek_n$$

where the conductivity is σ, the resistivity is ρ and k_n is the mobility of the electrons. Hence

$$k_n = 1/ne\rho = \frac{1}{10^{29} \times 1.6 \times 10^{-19} \times 1.5 \times 10^{-8}}$$

on substituting the values given and $e = 1.6 \times 10^{-19}$ C. Thus

$$k_n = 4.2 \times 10^{-3} \text{ m}^2 \text{ V}^{-1} \text{ s}^{-1}$$

1.14 Energy Levels in Isolated Atoms and Energy Bands in Solids; Metallic Conduction

In the gaseous state the mean separations (the mean free paths) between neighbouring atoms are several hundred times the diameters of the atoms themselves unless the gas pressure is many times greater than atmospheric pressure. The electrons around the nucleus, and even those farthest from the nucleus, in an atom in a gas are therefore relatively remote from those of the neighbouring atoms. Interactions between the outermost electronic structures of neighbouring atoms are negligible. From this point of view, the atoms (or molecules) in a gas may be said to be isolated from one another.

In the solid state, on the other hand, the separations between neighbouring atoms are comparable with the atomic diameters. Now the electronic structure of any one atom *will* interact with that of its neighbour atoms — the atoms are *not* isolated from one another.

Each and every electron in the extra-nuclear (outside the nucleus) structure of an isolated atom will occupy a discrete energy level, and these levels are identical with those of any other of the atoms. For the simplest atom — that of hydrogen — the main structure of the energy level diagram is shown in Figure 1.7. The data are obtainable by experiment (the study of the emission line spectra being the most useful technique) and is predicted with considerable accuracy by the theory of the hydrogen atom due to Niels Bohr (1913).

If atoms are close together in a solid, the atoms interact and the sharp electron energy levels (as represented for the hydrogen atom in Figure 1.7 with similar but more complex diagrams for the other isolated atoms) broaden into energy bands.

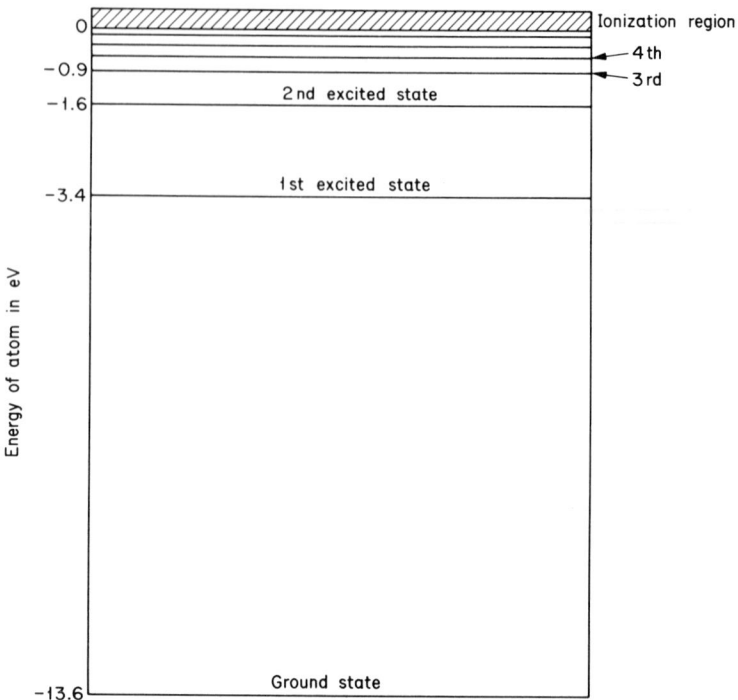

Figure 1.7. Energy-level diagram for the hydrogen atom

The simplest element which can exist at ordinary temperature in both the vapour and solid states is lithium (melting point 186°C). This has an atomic number Z of 3; there are two electrons in the filled K shell and one electron (the valence electron) in the L shell. For lithium vapour the sharp energy levels are shown in Figure 1.8(a). There are two electrons in the K shell with very nearly the same energy and the single valence electron in the L shell with a higher energy. The lithium atom can be excited (as in a discharge in lithium vapour) so the valence electron can be given temporarily still higher energy to occupy various discrete levels, normally unoccupied like that shown in Figure 1.8(a).

In solid lithium (actually a body-centre cubic crystal form), however, the sharp energy levels broaden into bands as shown in Figure 1.8(b). This broadening is due to the proximity of neighbouring atoms. Whereas in the isolated atom the electrons are under the electrostatic influence of only the nucleus of the atom about which they circulate, in the solid state the electrons are also under the electrostatic influences of the nuclei of the immediately neighbouring

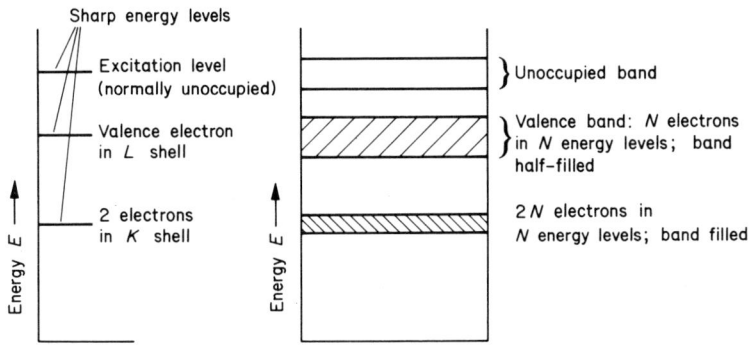

Figure 1.8. (a) Energy levels in isolated lithium atoms and (b) energy bands in a lithium crystal

atomic nuclei. Thus, the two valence electrons in a pair of neighbouring lithium atoms (considering the effect of only two very close atoms, for simplicity) may now occupy two energy levels, one slightly higher than that of the isolated atom and the other slightly lower. Whereas the single energy level in the isolated atom is E, there are now two energy levels at E_1 and E_2 spaced on either side of E. The value of $(E_1 - E_2)$ will increase the nearer the atoms are together.

Solid lithium has some 10^{28} atoms per cubic metre. There is the same number of valence electrons and twice this number of K shell electrons. Consequently there is an enormous number of energy levels which the electrons may occupy within certain upper and lower limits, depending on the proximity of packing of the atoms in the crystal. These innumerable energy levels form a continuous band of energies.

For lithium the proximity effect on the inner K shell electrons is less than for the valence electrons in the outer L shell. The lower energy band occupied by K shell electrons is therefore narrower than the higher energy band occupied by the L shell electrons. Again a still higher unoccupied band exists which electrons can attain if the lithium atoms are excited (Figure 1.8(b)).

Hence, in N atoms of lithium, where N is a very large number even for a small specimen, the set of energy levels is not the same for every atom (as it is in the gaseous state) but is different for every atom. There are, indeed, N possible energy states in the energy band. It can be shown that two electrons (with opposite spins about their own axes) may occupy the same energy level but not more than two. N energy levels in a band are thus filled when they contain $2N$ electrons.

In the case of lithium, the lower energy state (derived from the K shell electrons) is filled; the valence band, however, is only half filled because it can contain $2N$ electrons in its N levels whereas there are only N valence electrons within N lithium atoms.

If an electric field is applied to a lithium specimen, the valence electrons will accelerate and so have their energy increased, but only if there are energy levels to which they can be raised. Since half the levels in the band are available. electrons can gain this extra energy, i.e. electrons can move: conduction occurs. The same concepts apply to all the monovalent metallic conductors, in particular, sodium and potassium.

There are, of course, several other metallic conductors which are divalent, for example, beryllium, magnesium and calcium. A divalent metallic element will have two valence electrons per atom so the valence band will be filled. At first sight it would appear that energy cannot be given to these electrons on the application of an electric field. This is not the case because it turns out that these electrons can be excited into a conduction band which overlaps the valence band

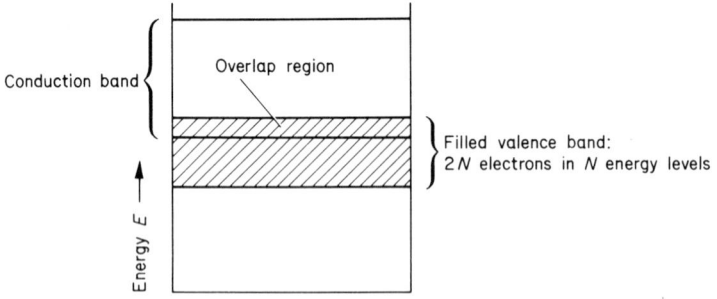

Figure 1.9. Energy band structure for a divalent metallic element where the conduction band overlaps the valence band

(Figure 1.9) and most of this conduction band (beyond the overlap region) is normally unfilled. The divalent metallic elements are therefore conductors because electrons can readily gain energy on the application of an electric field.

The energy band structures of all the metallic elements are such that conduction occurs because either the valence band is unfilled or the conduction and valence bands overlap.

1.15 Insulators

With the exception of sulphur and carbon in the form of diamond, the commonly encountered electrical insulators are compounds. A good example of an excellent insulator is aluminium oxide, Al_2O_3, two positive ions of aluminium are strongly bound by the three valence electrons per atom with three negative ions of divalent oxygen. The chemical bond energies are large, the valence electrons are firmly bound in the compound. None (or extremely few) are available for conduction in a specimen of alumina (the alternative name for aluminium oxide).

It is not an easy matter to deduce the energy band structures of insulators. However, they all have an important common feature. The valence band is filled and the higher unoccupied band into which the electrons could go is widely separated from the valence band (Figure 1.10). There is an *energy gap* of E_g electron-volt between the top of the valence band and the bottom of the nearest unoccupied level; E_g can typically be 9 eV.

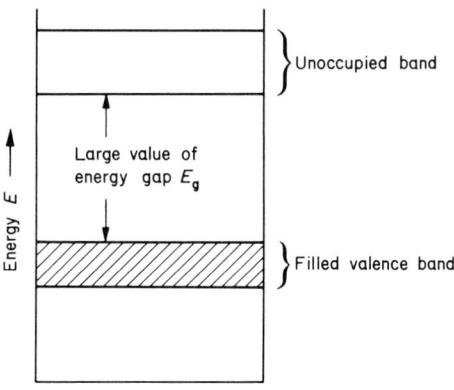

Figure 1.10. Typical energy band structure of an insulator

For the purposes of rough calculation, an electron at a temperature T K can be said to have a most probable energy of kT, where k is the Boltzmann constant. In fact the electrons will have a distribution of energies, but most of them will have energies of about kT. A very small fraction of the total will have energies of $10kT$ and a much small fraction still will have energies of $100kT$.

The Boltzmann constant $k = (1/11\,600)\,\text{eV K}^{-1}$. In order to acquire an energy of 9 eV, the temperature would have to be

$$T = 9 \times 11\,600 = 100\,000 \text{ K approx.}$$

This is meaningless because the insulator would vaporize. At only 1000 K, a very tiny fraction of the electrons would achieve energies of 9 eV, but so few that conduction would still be very small.

Note, however, that the conductivity of poor conductors (even though very small) increases with temperature rise, so insulators have a negative temperature coefficient of resistance.

1.16 Energy Band Structures of Semiconductors

As might be expected the energy band structure of the pure semiconducting element silicon is rather like that of insulators except that the energy gap between the valence and conduction bands is smaller.

To remove electrons from the covalent bonds of a pure semiconductor (to form an electron-hole pair) requires a significant amount of energy. The energy gap between the top of the valence band and the bottom of the conduction band is 1.1 eV for silicon.

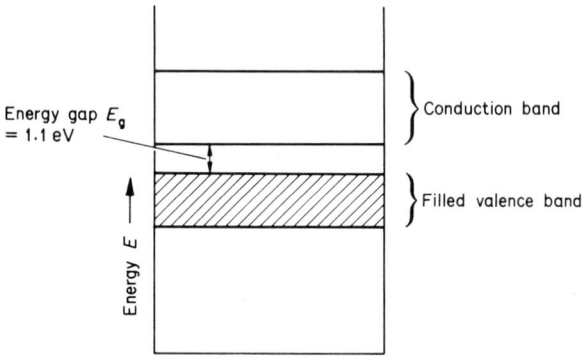

Figure 1.11. The valence and conduction band diagram for a silicon crystal

When a controlled amount of impurity is introduced into the lattice structure of a pure semiconductor it can become an impure semiconductor. Introducing a pentavalent impurity (to produce n-type silicon, for example) creates additional electrons to those in the covalent bonding and so creates new donor energy levels within the forbidden gap just below the conduction band (Figure 1.12(a)). The

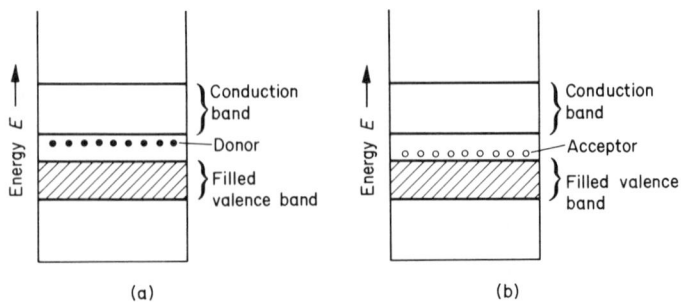

(a)   (b)

Figure 1.12. (a) Energy band diagram for a semiconductor doped with a pentavalent impurity and (b) energy band diagram for a semiconductor doped with a trivalent impurity

gap between this new level and the bottom of the conduction band is much less than the energy gap in the pure semiconductor. For silicon it is 1.1 eV; when doped with pentavalent phosphorus, for example, the new level is only, perhaps 0.05 eV or less below the bottom of the conduction band. Doped silicon is consequently a much superior conductor (with electrons as majority carriers) to pure silicon. Introducing a trivalent impurity (to produce p-type silicon, for example) creates additional positive holes and so produces new acceptor energy levels in the forbidden gap just above the top of the valence band (Figure 1.12(b)). Electrons from the valence band can readily be thermally excited into these vacancies leaving correspondingly vacant energy levels within the valence band which becomes partly unfilled. The conduction therefore increases greatly, with positive holes as the majority carriers.

Utilizing again a rough basis for calculation that an electron at a temperature T K has a most probable energy of $T/11\ 600$ eV, it is seen that at absolute zero (0 K) semiconductors will be insulators. As the temperature is increased, conductivity increases. For a pure semi-conductor such as silicon, the fraction of the electrons at $T = 300$ K which gain enough energy to surmount the gap of 1.1 eV is sufficient for conductivity to occur. When this gap is reduced to only 0.05 eV by the controlled addition of a suitable pentavalent impurity element, it is easily seen that, at a given temperature, the fraction of the electrons able to enter the conduction band is increased greatly. Furthermore, it is apparent that conductivity increases with temperature so that the temperature coefficients of resistance of these pure and impure semiconductors are negative.

If the semiconductor is heavily doped, most of the impurity atoms are ionized at room temperature. As the temperature is raised the mobility of the carriers decreases. Hence the materials show initially a positive temperature coefficient, similar to that of a metal. However, as the temperature is raised further, the number of electron-hole pairs produced more than compensates for this decrease in mobility and the resistance falls in the same way as that of a pure specimen. Such behaviour is shown in the region of 293 K in Figure 1.13 for a doped specimen of germanium.

1.17 An Experiment to Determine the Energy Gap E_g for Germanium

For near pure germanium the equation which connects the width of the forbidden band, that is the energy gap E_g, with the conductivity σ is

$$\sigma = B \exp(-E_g/2kT) \qquad (1.18)$$

where B is a constant having the dimensions of conductivity, T is the

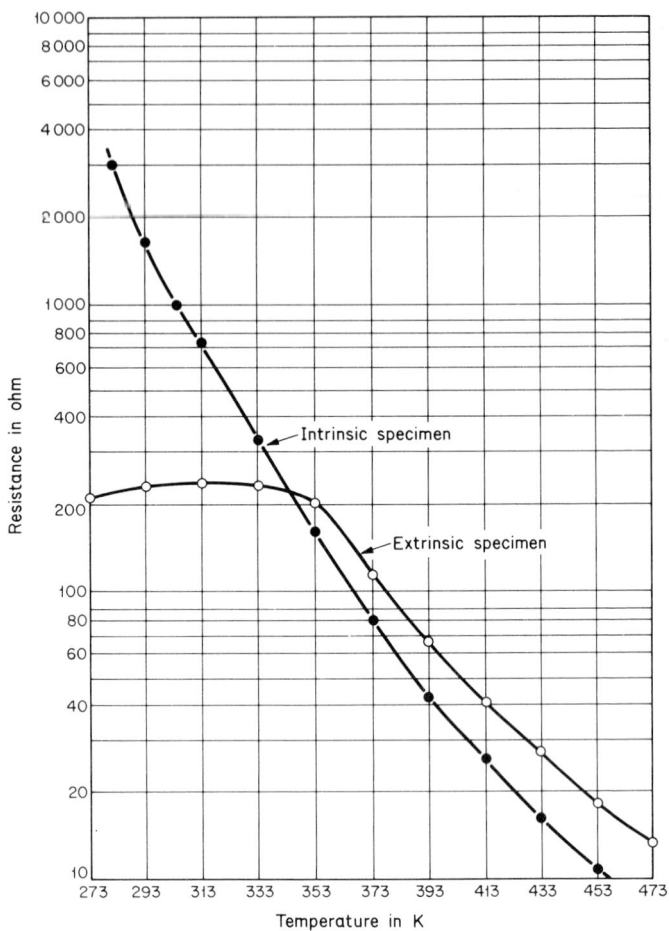

Figure 1.13. Plots of resistance against temperature for an intrinsic (near pure) and an extrinsic (doped) specimen of germanium

absolute temperature and k is the Boltzmann constant. The theoretical derivation of this equation is based on the energy band theory of solids. An experimental verification of this equation consequently gives support to the concept of energy bands. Moreover, it enables the value of E_g for germanium to be determined.

Equation (1.18) can clearly be expressed in the alternative form

$$\log_e \sigma = \log_e B - E_g/2kT \qquad (1.19)$$

If, therefore, the resistance of a uniform block of germanium of known dimensions is determined at various measured temperatures

T, σ is easily calculated and a plot of $\log_e \sigma$ against $1/T$ yields a straight line graph. The slope of this linear plot is $-E_g/2k$ so E_g is found.

A suitable experiment utilizes a short uniform rod of germanium about 15 mm in length, width 2 mm and thickness 2 mm. These dimensions are determined by a vernier or micrometer gauge. A constant current of 5 mA, say, is passed through this rod, which is provided with ohmic contacts at its ends.

This germanium rod is immersed in a beaker containing oil with the leads to the end contacts protruding above the surface. The oil is heated to various recorded temperatures (a mercury thermometer is used) by placing it on a gauze on a tripod beneath which is a suitable electrical heater. Maize oil obtainable from most grocers is suitable. Stirring of the oil is necessary to ensure a uniform temperature distribution.

At a number of recorded temperatures *T* within the range from about 20°C to 220°C, the potential difference *V* across the specimen is measured with a calibrated potentiometer or a high resistance voltmeter.

Providing that the current through the germanium rod is kept constant, σ is inversely proportional to the voltage *V*. Equation (1.19) therefore gives

$$\log_{10} V - \log_{10} C = 0.4343 \, E_g/2kT - \log_{10} B$$

where *C* is another constant. A graph of $\log_{10} V$ against $1/T$ therefore gives a straight line of slope $0.4343 \, E_g/2k$, as in Figure 1.14.

It is convenient to use a constant current source for the supply of about 5 mA, otherwise frequent adjustments of a series resistance in a simple battery supply circuit will be necessary as the temperature is altered. The circuit diagram of a suitable current source (also useful in other experiments) is described in section 4.11.

It is of interest to repeat this experiment with a doped specimen of germanium, e.g. n-type germanium containing antimony. Figure 1.13 shows the resistance plotted against the temperature for both a pure and an impure specimen.

The initial part of the curve for the doped specimen indicates a small positive temperature coefficient of resistance similar to that for a metal. In this temperature range up to about 300 K, all the impurity atoms are ionized (i.e. have lost an electron) and the number of electron-hole pairs produced does not compensate for the decrease in the mobility of the current carriers as the temperature rises. Hence the resistance of the specimen increases. At about 330 K, the intrinsic conductivity increases rapidly with temperature and so more than compensates for the decrease in mobility. At temperatures of about 350 K, the resistance change with temperature is similar to that for an

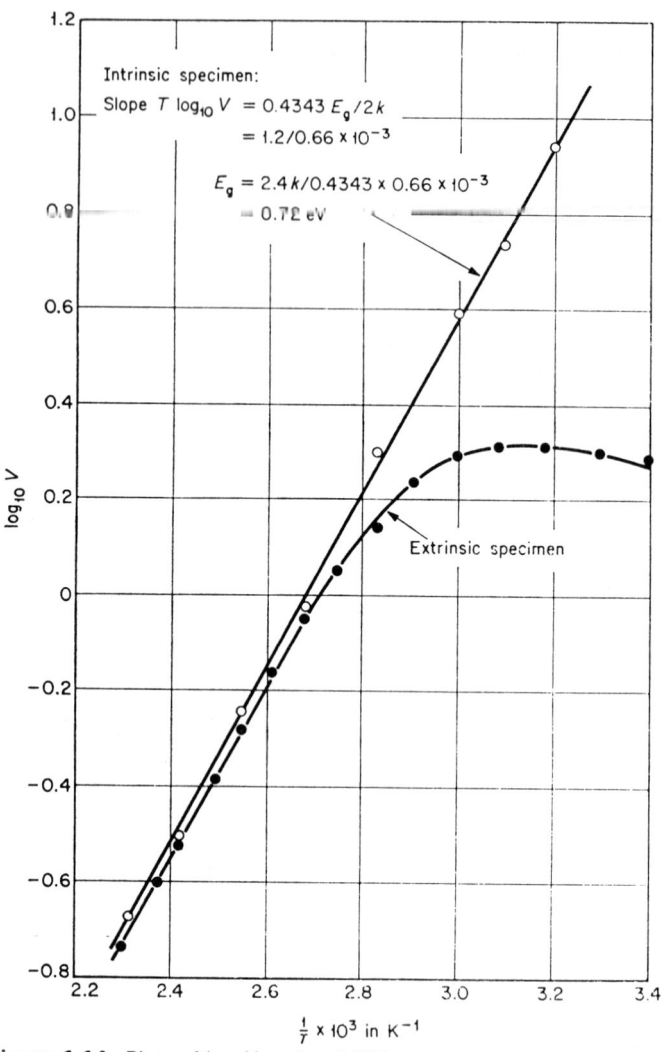

Figure 1.14. Plots of log V against $1/T$ for an intrinsic (near pure) and an extrinsic (doped) specimen of germanium

intrinsic specimen. Figure 1.14 shows plots of $\log_{10} V$ against $1/T$ for the two specimens each of which yields a value of 0.72 eV for E_g.

1.18 The Hall Effect

If a metallic or a semiconducting specimen carrying a current I is placed in a region in which the uniform magnetic flux density is B where the flux lines are directed perpendicularly to the current flow

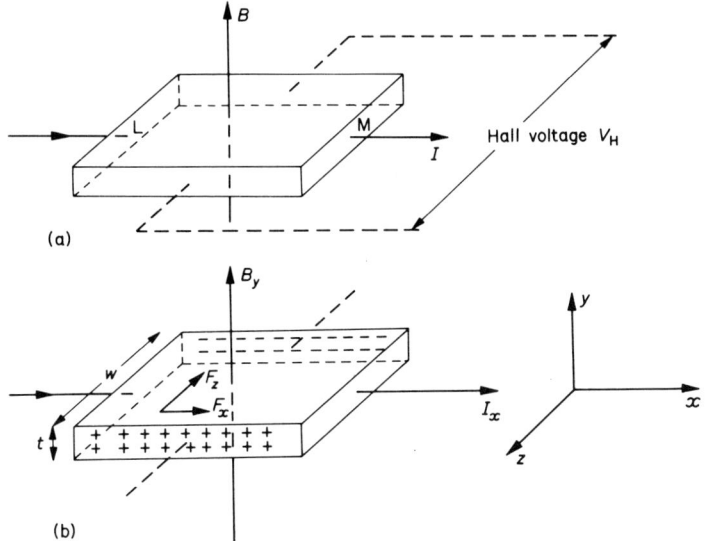

Figure 1.15. The Hall Effect

(Figure 1.15(a)), the current carriers within the specimen are deviated so that a voltage is produced across the specimen in the direction perpendicular to both B and I. This *transverse galvomagnetic effect* was first noted by E. H. Hall in 1879 and is known as the Hall effect.

In addition to the appearance of the Hall voltage V_H across the specimen, the deviation of the current carriers in the magnetic flux will cause a small change in the resistance between the faces L and M. This *transverse magnetoresistance* is related to the Hall effect and can be detected as a small current change when the magnetic flux density B is changed.

It is convenient in an experiment on the Hall effect to use the constant current source described in section 4.11 because frequent adjustments of the current through the specimen are avoided.

The Hall effect is much larger in semiconductors than in metals and is therefore easier to measure. It enables the majority carriers to be identified as either electrons or holes and also their concentration to be determined. Hence the material can be quickly identified as n-type or p-type and the impurity concentrations can easily be calculated.

The magnitude of the effect can be readily calculated for the motion of current carriers of charge q (which would be $-e$ for an electron and $+e$ for a positive hole).

Referring to Cartesian axes Ox, Oy, Oz (Figure 1.15(b)) let the current through the uniform specimen in the direction Ox be I_x. The current per unit cross-section area, i.e. the current density, J_x, is therefore given by

$$J_x = I_x/wt \qquad (1.20)$$

where w is the width and t the thickness of the specimen cross-section.

If E_x is the corresponding electric field strength across the specimen in the x direction, n is the carrier concentration (number of current carriers per unit volume) and k is the mobility of the current carriers,

$$J_x = nqkE_x \qquad (1.21)$$

The average force on a moving charge q due to the magnetic flux is in the z-direction (Fleming's left-hand rule) and is of magnitude

$$F_z = B_yqu$$

where B_y is the component of the magnetic flux density in the y-direction and u is the drift velocity in the x-direction. Now $J_x = nqu$. Hence

$$F_z = B_yqJ_x/nq$$

Using equation (1.21),

$$F_z = B_yqkE_x \qquad (1.22)$$

The deviation of charge by the magnetic flux will cause free charge to accumulate on the faces of the specimen perpendicular to the z-axis (Figure 1.15(b)). This charge will continue to accumulate until an equilibrium state is established. At this equilibrium, the force qE_z on the charge q due to the growing electric field E_z (produced by the free charge) will be equal and opposite to that force produced by the magnetic flux. Therefore

$$qE_z = F_z = qkE_x B_y$$

from equation (1.22). Therefore

$$E_z = kE_x B_y$$

Substituting for kE_x from equation (1.21),

$$E_z = J_xB_y/nq \qquad (1.23)$$

The *Hall coefficient* R_H is defined by

$$R_H = E_z/J_xB_y$$

Hence, from equation (1.23)

$$R_H = 1/nq \tag{1.24}$$

For a semiconductor, a more advanced theory gives

$$R_H = 3\pi/8nq \tag{1.25}$$

As n is the carrier concentration in number per cubic metre, and q, the charge, is in coulomb, the unit of R_H is seen from equation (1.24) or (1.25) to be metre cubed per coulomb ($m^3\ C^{-1}$). Normally, the Hall voltage due to the electric field component E_z is measured. This is given by

$$V_H = E_z w \tag{1.26}$$

Substituting into this equation (1.26) the value of E_z given by equation (1.23),

$$V_H = \frac{J_x B_y w}{nq}$$

which, from equation (1.20), becomes

$$V_H = \frac{I_x B_y}{tnq} \tag{1.27}$$

For a semiconductor (equation 1.25) this becomes

$$V_H = \frac{3\pi I_x B_y}{8tnq} \tag{1.28}$$

This Hall voltage is measured between metal probes attached to the opposite faces of the crystal specimen.

Thin wafers of n-type germanium (dimensions $5 \times 5 \times 0.4$ mm approx.) are suitable for Hall effect measurements. It is not easy, however, to make good electrical contacts on the crystal slice. A preferable experiment is therefore to use a Hall probe, which is a single crystal slice with four leads attached (Figure 1.16(a)). Then, instead of undertaking actual Hall effect measurements, one can establish experimentally that, in accordance with equation (1.28), V_H is proportional to I_x and also to B_y. Having calibrated the probe, a constant current is passed through from the type of source described in section 4.11, and it is used as a direct reading fluxmeter, making use of equation (1.28).

The four leads attached to the crystal slice (Figure 1.16(a)) are A and B to carry current and C and D for Hall voltage measurements. When

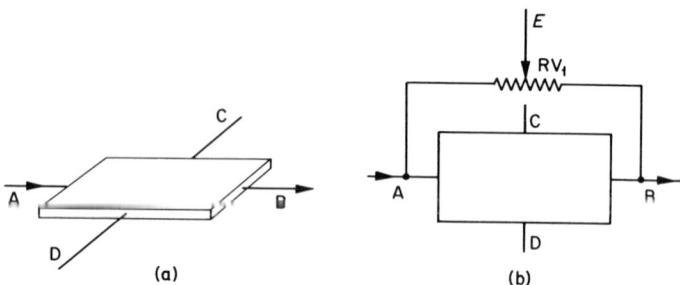

Figure 1.16. (a) A typical Hall probe and (b) use of a virtual contact to 'zero' the Hall probe

a current flows between A and B but no magnetix flux is applied, the voltage between C and D should be zero. For this to be the case, C and D must be located on an equipotential. It is not possible to arrange this. Consequently, the arrangement shown in Figure 1.16(b) is used. This provides a virtual contact which is often called a *virtual probe*, and effectively halves the specimen. With current flowing through the probe but no applied magnetix flux, the contact RV_1 is adjusted until the voltage between E and D is zero. The magnetic flux is then applied and the Hall voltage V_H is measured between E and D.

The identification of majority carriers by means of the Hall effect is possible. Suppose the current I (conventional direction from + to −) flows through a specimen (Figure 1.17). If the majority carriers are

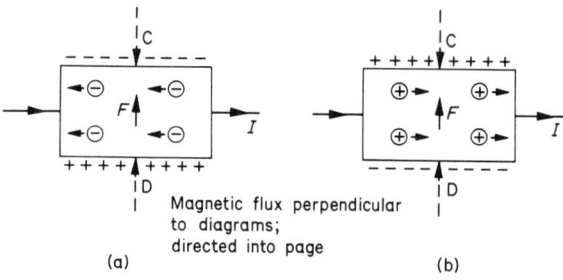

Figure 1.17. The identification of majority carriers by means of the Hall effect

electrons, they will move towards the left in Figure 1.17(a) and the force F on them upwards due to the transverse magnetic flux is of density B (Fleming's left-hand rule). Negative charge therefore accumulates on the top surface (contact C) and positive on the bottom surface (contact D). The voltage across CD is therefore with a certain polarity, D being positive with respect to C. When the majority carriers are positive holes, the direction of the force on them due to the magnetic flux is still the same. Now, however, the top surface becomes positively charged and the bottom surface negatively charged: the polarity of the voltage is reversed (Figure 1.17(b)).

Suppose an air-cored coil carries a current I_B. The magnetic flux density B_y produced within the coil is proportional to I_B. The Hall voltage V_H established across a semiconductor specimen within this flux is proportional both to the current I_x through this specimen and to B_y, i.e. to I_B. Hence

$$V_H \propto I_x I_B$$

A device of value in computers is thereby indicated in principle: it is capable of multiplying two quantities of magnitudes decided by I_x and I_B and where the product is proportional to V_H.

Example 1.18(a)

A copper strip carrying a current of 3 A is placed in a uniform magnetic flux of density 1.5 tesla with the flux lines at right angles to the direction of the current flow. If the width of the strip is 10.0 mm and its thickness is 0.15 mm, calculate the Hall voltage that appears across the strip if the number of copper atoms per cubic metre is 9×10^{28}. Assume that each atom contributes one free electron. (The electronic charge $e = 1.6 \times 10^{-19}$ C.)

From Equation (1.27)

$$V_H = \frac{I_x B_y}{tnq} = \frac{3 \times 1.5}{1.5 \times 10^{-4} \times 9 \times 10^{28} \times 1.6 \times 10^{-19}}$$

$$= \frac{4.5}{21.6} \times 10^{-5} = 2.1 \mu V.$$

Note that this voltage is very small even though the strip is carrying a current of 3 A in a high magnetic flux density. As a result Hall voltage measurements on metals are not usually attempted in school or college laboratories.

Example 1.18(b)

A rod of n-type germanium is 2.0 mm wide and 0.15 mm thick. A current of 10.0 mA is passed along its length and a uniform magnetic flux of density 0.2 tesla is established at right angles to the current flow. The Hall voltage is 2.5 mV. Calculate the Hall coefficient and the electron concentration.

For a semiconductor equation (1.28) applies:

$$V_H = \frac{3\pi I_x B_y}{8tnq}$$

The electron concentration is therefore

$$n = \frac{3\pi I_x B_y}{8 V_H e t}$$

because $q = e$, the electronic charge. Substituting the values given together with $e = 1.6 \times 10^{-19}$ C,

$$n = \frac{3\pi \times 10^{-2} \times 0.2}{8 \times 2.5 \times 10^{-3} \times 1.6 \times 10^{-19} \times 1.5 \times 10^{-4}} m^{-3}$$

$$= 3.9 \times 10^{22} \text{ m}^{-3}$$

From equation (1.25),

$$R_H = \frac{3\pi}{8nq} = \frac{3\pi}{8 \times 3.9 \times 10^{22} \times 1.6 \times 10^{-19}} m^3 \text{ C}^{-1}$$

$$= 1.9 \times 10^{-4} \text{ m}^3 \text{ C}^{-1}.$$

Notice the comparatively large Hall voltage V_H quoted for this n-type germanium sample even though a modest current and magnetix flux density are used. A sample such as this one is ideal for simple experimental work.

In the following exercise the abbreviation A.E.B. denotes that the question which it terminates has been taken with permission from papers set in Advanced level examinations by the Associated Examining Board, either in Physics or in the endorsement paper on Electronics.

Exercise 1

1. What information regarding the forces between molecules or atoms can be inferred from the physical properties of materials in the solid, liquid and gaseous states?

   Explain in terms of molecular motion why liquids cool as they evaporate.

2. Explain the terms *amorphous*, *crystalline*, *polycrystalline* and *isotropic*.

3. Calculate the speed of an electron which has been accelerated from rest through a potential difference of 3500 V. (Specific charge of electron, e/m_e, is 1.76×10^{11} C kg^{-1}.)

4. Calculate the root mean square speed of molecules of oxygen in a container at a temperature of 300 K and a pressure of 1.013×10^5 N m^{-2}, given that the density of oxygen at s.t.p. is 1.62 kg m^{-3}.

   An oxygen molecule is allowed to fall freely through a distance of 1.0 m. Calculate the energy acquired by this molecule and express this as a fraction of the average kinetic energy of an oxygen molecule at 300 K. (Molecular mass of oxygen = 32.00; Avogadro constant = 6.023×10^{26} per kilomole; gravitational acceleration = 9.8 m s^{-2}; Boltzmann constant = 1.38×10^{-23} J K^{-1}.)

5. Electrons which have been accelerated from rest through a potential difference of 2.0 kV are injected into a uniform magnetic flux of density 2×10^{-3} tesla in a direction at right angles to the flux lines. Calculate the radius of curvature of the electron trajectory. (Specific charge of electron, e/m_e, is 1.76×10^{11} C kg^{-1}.)

6. 'Free electrons in a metallic conductor behave in a similar fashion to the molecules of an ideal gas.'

    Discuss the behaviour of electrons in metals which supports this statement.

7. Outline aspects of the behaviour of the elements sodium, carbon and silicon that can be inferred from the knowledge that their atomic numbers are 11, 6 and 14 respectively.

8. Compare and contrast the electrical behaviour of metals and semi-conductors.

9. Explain the classification of materials into conductors, insulators and semiconductors in terms of electron energy levels.   (A.E.B.)

10. The atomic weight of sodium (Na) is 22.99 and that of chlorine (Cl) is 35.46. If the density of a large crystal of rock salt (NaCl) is 2.163×10^3 kg m^{-3}, calculate its lattice constant (the separation between neighbouring parallel planes of atoms). (Avogadro constant $= 6.025 \times 10^{26}$ per kilomole.)

11. Discuss the electronic structure of atoms of elements which may be used to dope a specimen of silicon to produce (*a*) n-type silicon and (*b*) p-type silicon.

    How does the process of doping modify the energy band structure of a semiconductor?

12. Write short notes on
    (i) intrinsic and extrinsic conductivity;
    (ii) majority and minority carriers;
    (iii) carrier mobility.   (A.E.B.)

13. What is meant by (*a*) intrinsic, and (*b*) extrinsic conductivity?

    Explain how and why the electrical conductivity of a crystal slice of germanium varies with:
    (*a*) temperature over the range 273–523 K;
    (*b*) trivalent impurity atoms present;
    (*c*) pentavalent impurity atoms present.   (A.E.B.)

14. The following two experiments are undertaken:
    (*a*) the resistance of a rod of intrinsic germanium is measured as its temperature is raised from 300 K to 500 K;
    (*b*) the resistance or a rod of n-type germanium is measured as its temperature is raised from 200 K to 500 K.

    Sketch for both cases the curve of resistance against temperature which would be obtained and explain the processes which determine the shapes of these curves.

    How could the energy gap (the width of the forbidden band) for germanium be determined from either set of these experimental results?

15. Derive an expression for the conductivity of a semiconducting material in terms of the concentration n and the mobility k of the electrons and holes.

    A rod of intrinsic germanium of dimensions 10.0 mm $\times$ 2.0 mm $\times$ 1.0 mm has a resistance of 150 Ω at a temperature of 400 K. If the electron and hole mobilities are respectively 0.37 m^2 V^{-1} s^{-1} and 0.18 m^2 V^{-1} s^{-1}, calculate the electron concentration in the specimen at this temperature. (Electronic charge $e = 1.6 \times 10^{-19}$ C.)

16. An intrinsic specimen of germanium at 300 K has a resistivity of 0.47 Ω m. Calculate the concentration of electrons if the electron and hole mobilities are respectively 0.36 m² V^{-1} s^{-1} and 0.17 m² V^{-1} s^{-1}.

17. For germanium at a certain temperature it is known that the number of electrons per unit volume and of holes taking part in intrinsic conductivity is 3.0×10^{19} m^{-3}. At this same temperature, a rod of p-type germanium of dimensions 10 mm × 2 mm × 0.5 mm has a resistance of 100 Ω. Calculate the impurity concentration in this specimen assuming that all the acceptor atoms are ionized. (Electron charge $e = 1.6 \times 10^{-19}$ C; electron mobility $k_n = 0.39$ m² V^{-1} s^{-1}; hole mobility $k_p = 0.19$ m² V^{-1} s^{-1}.)

18. What is the Hall effect? Explain how Hall voltage measurements on a semiconducting crystal slice can be used to identify the majority carriers and to determine their concentration.

2 The manufacture of semiconductor devices

Although a very large number of semiconducting materials is known, only a relatively small number is at present being used to make electronic components. Germanium and silicon in single-crystal form are by far the most common elements used but the search continues in laboratories throughout the world for other suitable materials. Whether in single-crystal or polycrystalline form, the aim is to provide cheaper, more reliable and more versatile semiconductor devices.

In the past decade silicon has replaced germanium in most components and this preference for silicon has become paramount with the advent of integrated circuits. Germanium was used first to make most of the transistors because it is easier to purify. Subsequently, silicon has been proved to have superior characteristics. In particular, a silicon component can operate over the wide temperature range from $-50°C$ to $150°C$ with negligible deterioration in its electrical behaviour.

The problem which has faced the semiconductor industry appears simple: to produce a junction between p- and n-type materials to make a rectifier, and two such junctions in a single component to make a junction transistor. The solution has nevertheless proved very difficult and has demanded an almost unprecedented effort to provide semiconducting elements of adequate purity with controlled additions of different known elements, to manufacture junction transistors with exceedingly close p-n junctions, and provide known crystal structures with reliable and reproducible electrical characteristics.

In this unique industrial development chemists contribute by providing methods of extracting and purifying the elements and investigating reactions which clean, etch and stabilize the crystal surface. Physicists explain on a basis of solid-state physics the behaviour of existing components and provide, in conjunction with electronic engineers, interesting new designs and circuit applications. Metallurgists study the production of alloys and the techniques of growing large single crystals while mechanical engineers develop

43

intricate machines for slicing, handling and encapsulating miniature components and assemblies of components in integrated circuits (i.c.). Often, the environment under which these semiconductor devices and circuits are assembled is more akin to a hospital than a factory. Indeed, in the manufacture of integrated circuits attention to both process control and clean conditions is absolutely essential because it is imperative to avoid defects (e.g. due to dust particles) when several thousand components, with the necessary interconnections and terminals, may be fabricated within a chip of silicon which is only 6.25 mm square, or less.

2.1 The Zone Refining of Germanium and Silicon

Germanium is in group IVA of the periodic table which comprises carbon, silicon, germanium, tin and lead, all quadravalent elements of which the atomic numbers are respectively 6, 14, 32, 50 and 82. Germanium is a rare element with a grey metallic appearance and is generally extracted from sulphide ores of zinc, lead and copper in which it occurs in low concentration. It is also found in the ashes of certain coals.

Silicon is the second most abundant element in the earth's surface but usually occurs in combination with oxygen as silica (SiO_2) and is difficult to prepare in a pure state.

For the semiconductor industry, both silicon and germanium are obtained in a pure state from the tetrachloride or the dioxide. For the manufacture of semiconductor diodes and transistors, however, the demands on purity are considerably greater than is usually encountered in the most refined analytical chemistry. Whereas an impurity concentration of 1 part in 10^6 is considered top class work in normal chemical practice, only about 1 foreign atom in 10^9 atoms can be tolerated in germanium or silicon suitable for semiconductor devices.

The zone-refining method using a horizontal furnace as in Figure 2.1 is suitable for germanium (melting point: 958°C), but for silicon (melting point: 1420°C) a *floating zone process* is preferable in which the bar of silicon is clamped in a vertical position in a quartz tube. A molten zone, containing most of the impurities is made to traverse the bar, the molten region being held in position by surface tension forces. Very careful temperature control is demanded. A doping material added at one end can be uniformly distributed along the bar by means of a single cycle of zone refining.

2.2 Growing Large Single Crystals

One method often used to grow large single crystals of germanium or silicon for the manufacture of transistors is due to Czochralski (Figure 2.2). Within an inert atmosphere an electrically heated

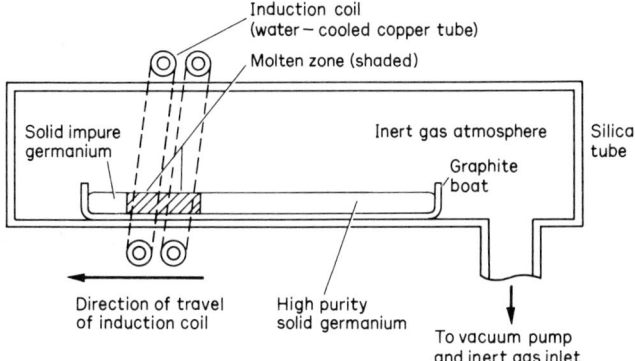

Figure 2.1. The technique of zone refining

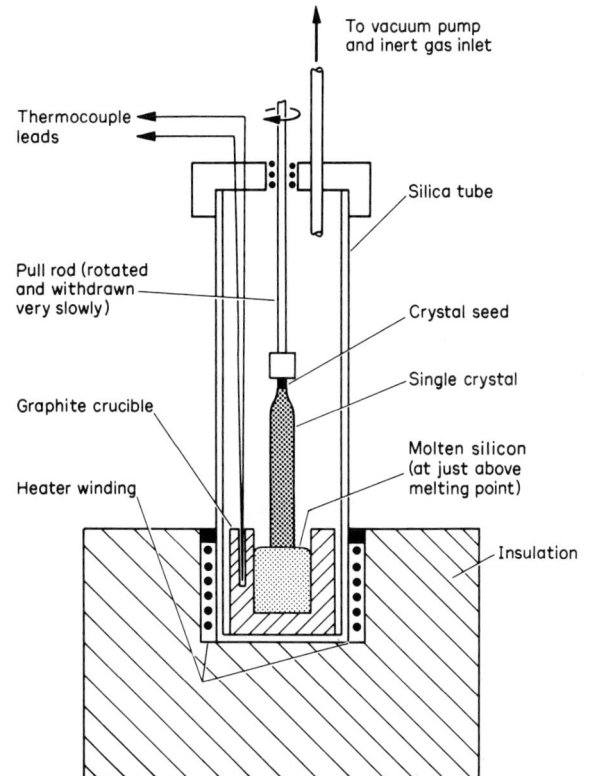

Figure 2.2. The Czochralski method of growing single crystals

crucible maintains the silicon at a temperature just a few degrees above its melting point. A very small piece of single crystal, called a *seed*, is lowered on to the surface of the melt and then slowly withdrawn. Surface tension forces support some of the molten material, which cools slowly and solidifies with an orderly arrangement of atoms identical with that of the seed. Over a period of perhaps twenty-four hours, a single crystal 50 mm in diameter* and 300 mm long could be grown.

Trivalent or pentavalent doping elements can be added to the melt before the crystal is grown to produce a p- or n-type crystal of known resistivity. This resistivity, governed by the impurity content, will depend on the rate at which the impurities are transferred from the melt to the crystal and can be controlled by varying the crystal growth rate.

2.3 Forming a p-n Junction

A junction is ideally a surface separating a p-type semiconductor from an n-type. The very useful electrical properties of such a junction cannot be reliably obtained by placing a p-type material in contact with an n-type. Indeed, the transition must be produced within a single crystal slice.

In some components the transition needs to be as sharp as possible; in others, the transition needs to extend over a finite region of the crystal. These two extremes are referred to as *step* and *graded* junctions respectively. Of the many processes used to produce a sharp change in the impurity concentration within a semiconductor crystal lattice, only diffusion will be described in some detail.

The semiconductor is heated in a vapour of the impurity atoms which enter the crystal surface and diffuse slowly into the crystal lattice. The depth to which the impurity atoms penetrate is governed by the temperature and the time. This diffusion process normally takes several hours and temperatures in the range 500°C to 600°C are used for germanium and 900°C to 1300°C for silicon.

2.4 Junctions Produced by Diffusion

The production of p-n junctions by diffusion is of considerable importance in transistor technology. Diffusion methods make it possible to control very precisely the concentration and the concentration gradient of an impurity over very small regions of a semiconductor. As an example of this process, the construction of a silicon planar diode is described.

The planar process comprises three parts:

---

* Single crystals of silicon of as much as 100 mm (or even 125 mm) in diameter are now grown.

(*a*) Covering a slice of n-type silicon with a hard inert oxide layer which acts as a diffusion mask.

(*b*) Etching holes in the oxide layer to expose and define those areas of silicon into which impurity may diffuse.

(*c*) Diffusing in from the vapour phase the trivalent impurity, which is usually boron.

Steps in the construction process are shown in Figure 2.3. From a slice of n-type silicon approximately 50 mm in diameter and 0.15 mm thick, about 4000 low-power diodes can be constructed. A layer of

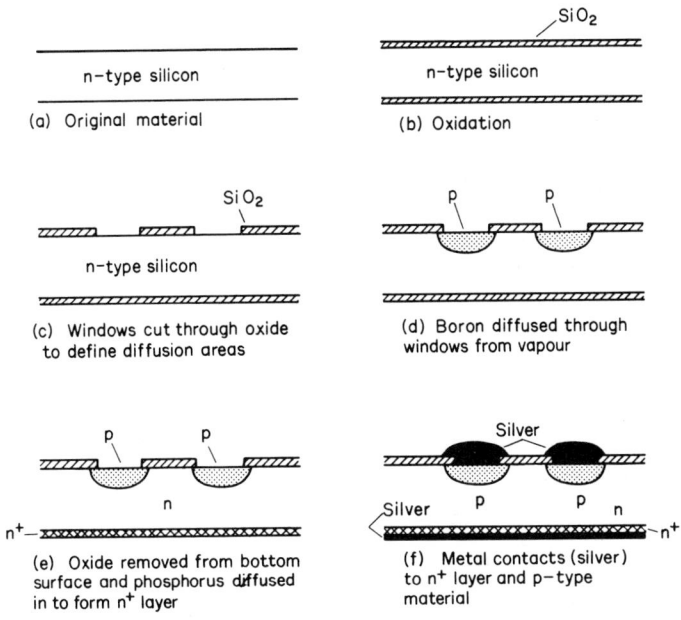

Figure 2.3. The diffusion process in making a silicon planar diode

oxide (SiO$_2$) is formed on the slice by heating it to high temperatures in a steam atmosphere. Holes are cut in the oxide layer to expose the silicon. In another high temperature tubular furnace, boron is diffused from the vapour into the exposed areas of silicon to form a p-n junction. The oxide which covers the back surface of this slice is now removed and pentavalent phosphorus is diffused into the back in a third high temperature furnace. This very highly doped n region is normally termed an n$^+$ region, and is so labelled in Figure 2.3. This term can be misleading: it is merely used to signify that the region is doped very heavily, has a conductivity similar to that of a metal and allows an ohmic (non-rectifying) contact to be made to it without difficulty.

The centre of the p-type region is exposed and silver or gold contacts are made, one to the p-region and the other to the n^+ region.

The planar diode has an oxide covering except where the contacts are exposed. This inert layer eliminates the very troublesome electrical behaviour of a contaminated surface. The surface is said to be *passivated* and the diode is known as a *passivated planar diode*. This diffusion process with the oxide masking can be employed to create transistors of either the bipolar junction or the field effect type (both are described in Chapter 4) with well-controlled characteristics. In addition, the technique allows a complete electronic circuit to be constructed in the face of a single chip of single crystal silicon. This circuit, called an *integrated circuit*, when mounted may be only a little bigger than a single transistor. Microcircuits of this kind — the foundation of the new technology of microelectronics — are very fruitful and provide enormous savings in space, cost and weight in space vehicles and in the ever more complex computer. In this latter connection, it must suffice here to state that the use of integrated circuits has enabled digital computers to be made today which are at least 20 times faster and much more reliable than those of 20 years ago which were built around large numbers of discrete transistors and passive components (capacitors and resistors).

The fabrication of integrated circuits is described in Chapter 6. Although the integrated circuit has now taken over in most applications from the discrete transistor, it is considered important to understand these components as discrete devices before their production within integrated circuits is described.

Exercise 2

1. Explain, with the aid of a diagram, the zone-refining process used to obtain silicon in a very pure state.
2. A contact between a metal and a semiconductor is specified as an ohmic contact. Explain the meaning of this term and outline one method used to produce such a contact.
3. Describe, with a diagram of the apparatus, a method of growing large single crystals of silicon.
4. Describe, with diagrams, the process of forming a p-n junction in silicon by diffusion.

3 Semiconductor diodes and rectification

3.1 Power Supply

The operation of electronic apparatus almost always requires power supplies. The familiar example is the cathode ray tube which needs a low voltage (low tension, L.T.) supply for its filament or cathode and a high voltage (high tension, H.T.) supply across its anode and cathode. Semiconductor devices, such as the transistor, offer immediate practical advantages in this respect: there is no filament or cathode heater and the supplies needed are normally at much lower voltages than those required for thermionic vacuum tubes: in most cases only a few volts is required as against 100 volts or more.

Dry batteries such as the familiar six-cell 9 volt ones, are often used as power supplies for semiconductor electronic circuits, the well-known example being the transistor radio set; another instance is the electronic measuring instrument, for example, the digital voltmeter.

There is, however, the need to provide a steady voltage supply (a so-called d.c. voltage) by making use of the a.c. mains. The a.c. mains supply is commonly at 240 V r.m.s. and alternates at a frequency of 50 Hz. This voltage is frequently too high or too low for the practical purpose required. It is therefore first changed to a different value by means of an a.c. transformer. This transformer is of the step-down type ($T < 1$ where T is total number of turns on the secondary winding divided by the total number on the primary winding) if the output r.m.s. voltage is to be less than 240 V, and of the step-up variety ($T > 1$) if the output is to exceed 240 V r.m.s.

This transformer has the desirable feature also that the output is isolated, as regards direct electrical connection, from the mains.

The secondary voltage is alternating: its output varies sinusoidally with time (Figure 3.1(a)). It is essential to render this output unidirectional to provide eventually a d.c. voltage. This means that the positive half-cycles, say, are needed, but the negative half-cycles have to be eliminated. The process whereby this is achieved is called *rectification*. A device called a *rectifier* is needed which conducts electricity readily when the voltage across it is in one direction but which conducts electricity poorly (ideally, not at all) when the

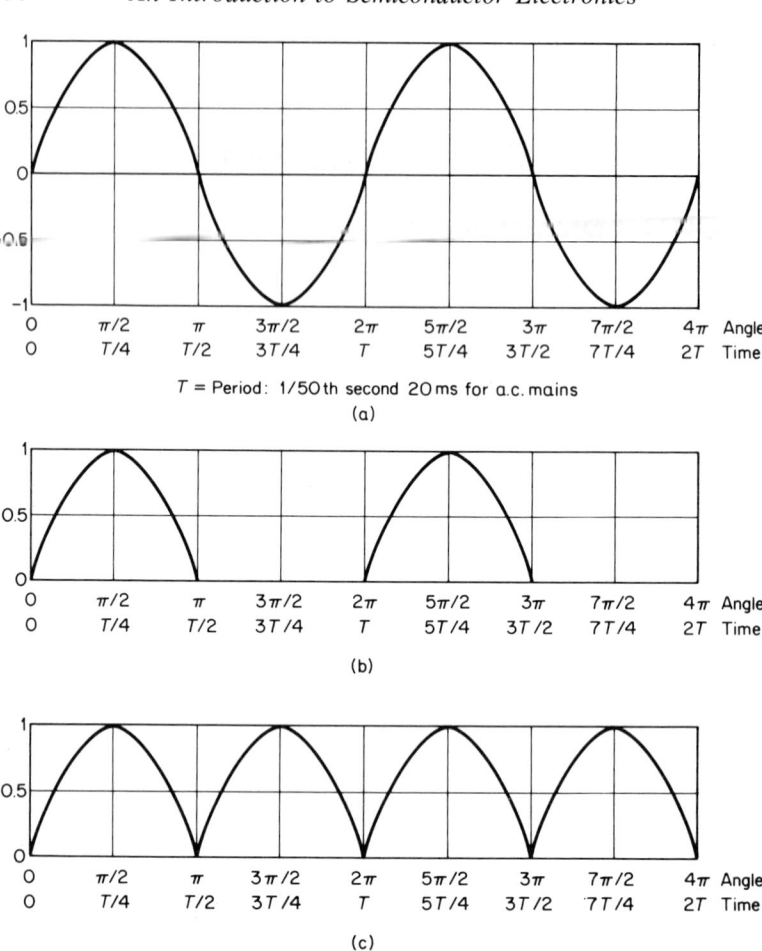

Figure 3.1. (a) The sine waveform of an a.c. supply, (b) half-wave rectification: the waveform, and (c) full-wave rectification: the waveform

voltage is reversed. The semiconductor diode is the modern and convenient rectifier.

Elimination of the half-cycles of one polarity (say the negative ones) results in a voltage of the waveform shown in Figure 3.1(b). This is known as *half-wave rectification*; if, by circuit ingenuity, the undesired half-cycles are not eliminated but reversed in polarity, *full-wave rectification* (Figure 3.1(c)) is performed.

The half-wave or full-wave rectified supply is not satisfactory for most power supply purposes. It cannot be envisaged that the voltage across the electronic apparatus is allowed to vary between zero and a peak value each cycle or twice a cycle. Imagine the unfortunate noise that would result from a radio set with such a power supply!

It is therefore necessary to smooth these wide fluctuations of voltage. This means that the voltage has to be constant and not vary with time. The graph of voltage against time would then be simply a straight line parallel to the time axis. This smoothing is achieved by capacitors or, better, a smoothing filter involving capacitors and resistors.

Though steady d.c. voltage supplies are hence obtained by rectification and smoothing, it is frequently the case in the operation of electronic apparatus used for measurement purposes that the voltage is not sufficiently constant. Thus, it might be constant to within ±1 per cent but the demand is that it be constant to within ±0.001 per cent or better. This additional requirement is met by a *voltage stabilizing circuit*.

The most widely used rectifier is the silicon p-n junction, known as the *silicon diode*. The simplest stabilizer is a silicon reference voltage p-n junction, a *Zener diode*. Their use has revolutionized power-pack design.

Stabilization of current is also a requirement as well as stabilization of voltage. Zener diodes as reference voltage devices are described in this chapter. The more complex circuits capable of maintaining either the voltage or the current at a fixed value will be examined in Chapter 4.

Apart from their obvious advantages of size and weight compared with rectifier valves and the old metal rectifiers, silicon diodes have proved to be very efficient and reliable where high power is demanded as well as for the small power requirements of small-scale electronic apparatus.

In Figure 3.2(a) is shown a selection of semiconductor diodes; included is a silicon diode capable of passing a current of 20 A. Although the mounting and cooling of this high-power component is a separate problem from the concerns of the small power apparatus described in this text, yet its electrical characteristics are essentially similar to those of the modest diodes.

Transistor outlines (TO) have internationally agreed sizes. Two are TO5 and TO18 of which typical dimensions in mm are tabulated below for Figure 3.2(b):

	A	B	C	D	E	F	G	H	J
TO5	9	8	6	5	0.8	40	0.5	0.4	0.9
TO18	6	5	5	3	1	13	0.4	0.05	1

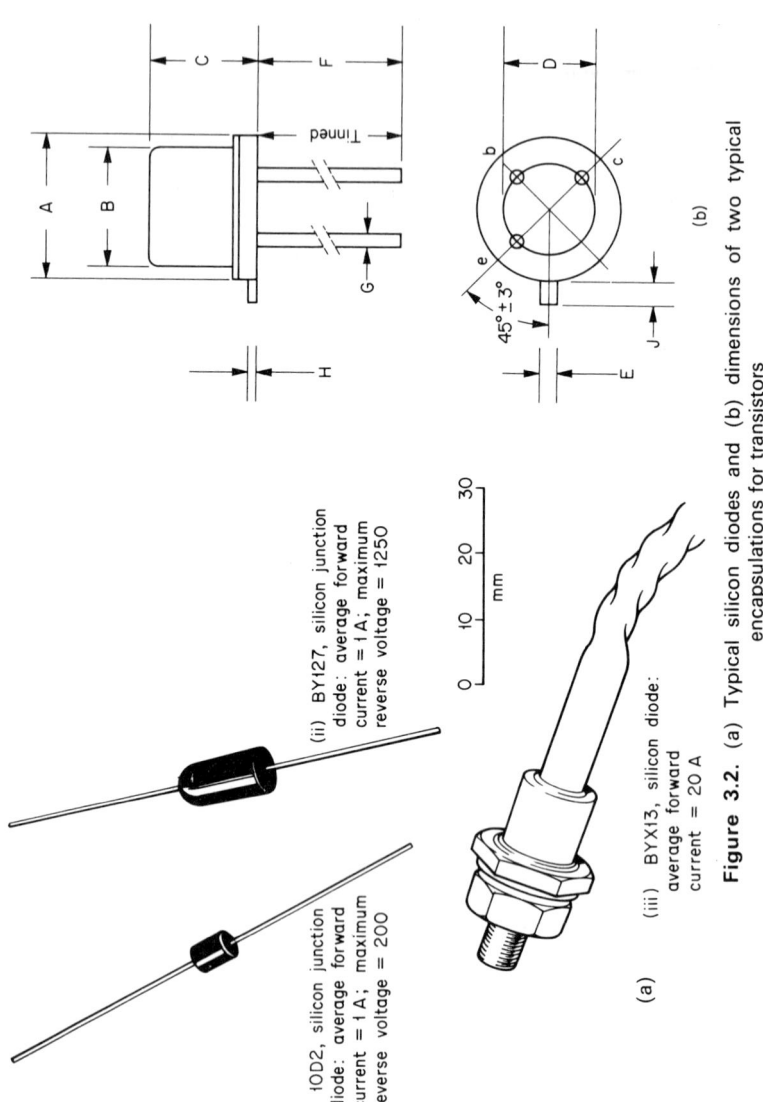

(i) 1OD2, silicon junction diode: average forward current = 1 A; maximum reverse voltage = 200

(ii) BY127, silicon junction diode: average forward current = 1 A; maximum reverse voltage = 1250

(iii) BYX13, silicon diode: average forward current = 20 A

(a)

(b)

Figure 3.2. (a) Typical silicon diodes and (b) dimensions of two typical encapsulations for transistors

3.2 The Electrical Characteristics of the p-n Junction Diode

The diode, a two-terminal device, offers negligible resistance to the flow of current in one direction and yet presents a nearly infinite resistance in the opposite direction. The circuit symbol (Figure 3.3(a)), which is often printed on the component case (Figure 3.3(b)), shows how the device should be connected for a particular behaviour.

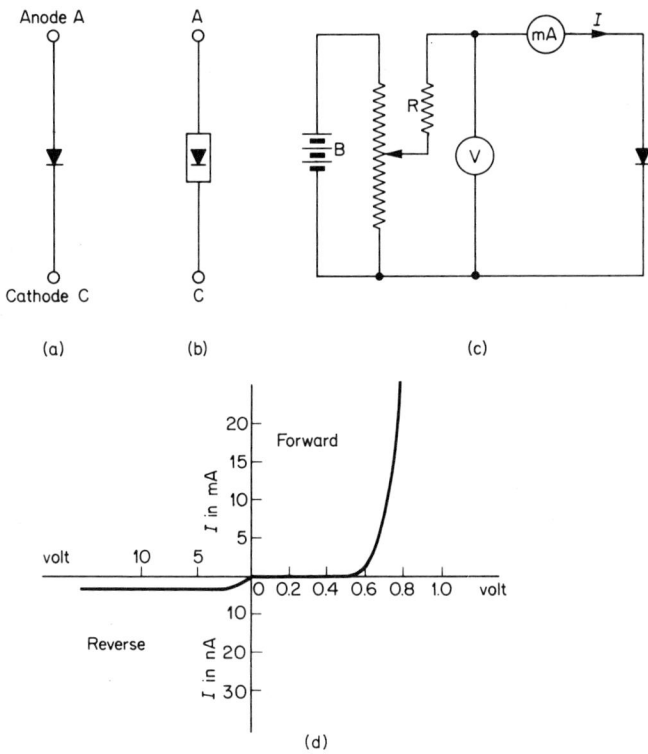

Figure 3.3. The silicon diode: (a) circuit symbol; (b) the component; (c) circuit for obtaining current against voltage characteristic; and (d) the current-voltage characteristic

The conventional current direction is from positive to negative (that in which a positive charge would move); the electron current is the other way, from negative to positive. Conventional current flows easily in the direction indicated by the arrow (Figure 3.3(a)).

When the terminal connected to the arrow head (the anode) is positive with respect to the other terminal, represented by a short line against the point of the arrow head (the cathode), the diode is said to be *forward biased.* When the diode is forward biased, conventional current flows readily in the direction of the arrow; electrons flow in the opposite direction from cathode to anode.

If the polarity of the voltage across the diode is reversed (arrow head negative; other terminal positive) the diode is *reverse biased.* Then only a small leakage current of a few nanoamperes will flow.

Sometimes, instead of the circuit symbol being printed on the component casing, a band is marked around this casing at the cathode end.

To determine the current through the silicon diode for various potential differences across it, the circuit shown in Figure 3.3(c) is used. The experiment with this circuit arrangement is in two parts. In part (i) the forward biased characteristic is obtained; in part (ii) the reverse biased.

For part (i) it is necessary to include a 50 Ω resistor R in series in the circuit to limit the forward current; otherwise the diode will be damaged if 1.5 V is placed directly across it. The maximum p.d. applied is 1.5 V and the moving-coil voltmeter V should be capable of reading 0 to 1.5 V in 0.1 V steps. The milliammeter mA has a full-scale deflection of perhaps 50 mA.

The current I recorded by mA is recorded for values of the p.d. V across the diode from zero in steps of 0.1 V up to 1.5 V. The plot of the forward characteristic is typically as shown in Figure 3.3(d).

To undertake part (ii) of the experiment, *the polarity of the battery B is reversed*: the arrow head (anode) is now made negative with respect to the cathode. The p.d. across the diode to obtain the reverse bias characteristic needs to be considerably larger than in part (i). The e.m.f. of the battery B is now about 10V. The voltmeter V has to be changed for one with a scale reading from 0 to 10 V in 0.5 V steps (or the range of a multirange voltmeter is increased). The reverse current is only a fraction of a nanoampere corresponding to a reverse resistance of several megohm. The meter mA hence has to be changed to a nanoammeter, if one is available.

Note that the vertical axis of Figure 3.3(d) is marked in mA in the positive direction and nA in the negative direction.

The current I which flows through a p-n junction at an absolute temperature of T when a potential difference V is maintained across it is given quite accurately by the equation

$$I = I_0[\exp(eV/kT) - 1] \tag{3.1}$$

where I_0 is a constant current of value about 5 nA (5×10^{-9} A) for a silicon p-n junction, k is the Boltzmann constant and e is the electronic charge. As $e = 1.6 \times 10^{-19}$ coulomb and $k = 1.38 \times 10^{-23}$

joule deg^{-1} K, so at $T = 300$ K $(27°C)$

$$\frac{e}{kT} = \frac{1.6 \times 10^{-19}}{1.38 \times 10^{-23} \times 300} \text{coulomb per joule}$$

$$= 38.5 \text{ per volt} = 38.5 \text{ V}^{-1}$$

Equation (3.1) may therefore be written

$$I = I_0[\exp(38.5V) - 1] \tag{3.2}$$

where V is the numerical value of the p.d. in volt across the diode. This voltage is positive when the junction is forward biased and negative when it is reverse biased. With reverse bias, put $V = -1$ V in equation (3.2), then

$$I = I_0[\exp(-38.5) - 1]$$

Here $\exp(-38.5) = e^{-38.5} = 1/e^{38.5}$ is negligibly small compared with 1 because $e = 2.71828$. The current I is therefore $= I_0$.

Consequently, with a reverse bias exceeding about 1 V, the reverse current is constant at I_0, which is about 5 nA (Figure 3.3(d)). I_0 is known as the *reverse saturation current*. This reverse saturation or leakage current increases with temperature T. Nevertheless, silicon diodes can be operated with a performance only slightly below their optimum at temperatures up to 150°C.

In the forward direction, V is positive in equation (3.2). For the forward current to become a few milliampere, $\exp(38.5V)$ in equation (3.2) must become large compared with unity. With $I_0 = 5$ nA, the forward current I will be 1 mA approximately at $T = 300°C$ when

$$\exp(38.5V) = 1 \text{ mA}/5 \text{ nA} = 2 \times 10^5$$

Hence

$$38.5V = \log_e 2 \times 10^5$$

$$= 2.3 \log_{10} 2 \times 10^5$$

$$= 2.3 \times 5.3010 = 12.19$$

$$V = \frac{12.19}{38.5} = 0.32 \text{ V}$$

It follows that the current in the forward direction is very small until the forward bias exceeds 0.3 to 0.7 V; above these values, this current increases exponentially as shown in Figure 3.3(d), its value being limited only by the series resistance R. Indeed, if R is omitted, the diode will be ruined because there is nothing to limit the forward current.

3.3 Explanation of the Electrical Behaviour of a p-n Junction Diode in Terms of its Structure

As soon as a junction is created between p-type silicon and n-type (indeed, for any p- and n-type materials), electrons which predominate in the n-type silicon will move across the junction to fill some of the preponderant positive holes (vacancies) in the p-type material Conversely, holes move across the junction in the opposite direction from the p-type to the n-type.

On the n-type side of the junction, loss of electrons from the material will leave some atoms with an excess positive charge, i.e. positive ions are created. On the p-type side, negative ions are produced. These ions are locked in the crystal lattice structure: they are not mobile like the electrons and holes. An electric field (due to these ions) will therefore exist across the p-n junction and directed from the n-type to the p-type material (Figure 3.4(a)). The field direction is the conventional one in which a positive charge would move or tend to move.

This motion of electrons and holes, leaving the unbalanced ion charges which produce the electric field, occurs instantaneously on forming the junction. The field direction is such as to oppose further electron or hole migration; an equilibrium is quickly attained.

For convenience in explanation, the inevitable field and

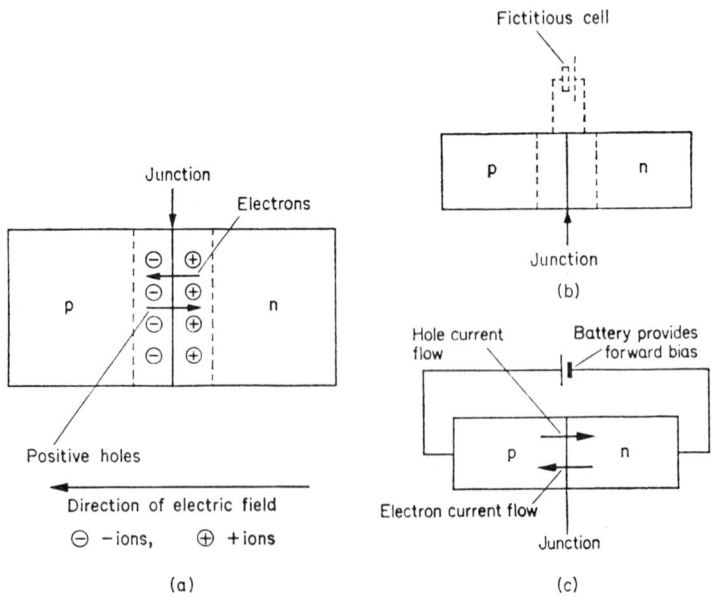

Figure 3.4. The electric field across a p-n junction

consequent p.d. (and remember that no external source of e.m.f. has yet been applied) is represented by a fictitious cell shown in dotted lines in Figure 3.4(b).

The region close to the p-n junction on both sides of it from which free current carriers have moved is called the *depletion region*.

When an external source of e.m.f. is applied across the p-n junction which opposes the e.m.f. of the fictitious cell, current will flow and the junction is forward biased (Figure 3.4(c)). This opposition corresponds to the positive terminal of the external voltage supply being connected to the p-type material and the negative terminal to the n-type.

Reversing the polarity of the externally applied e.m.f. will clearly increase the electric field across the p-n junction and no majority carriers (electrons in n-type, holes in p-type) can flow through the p-n junction. However, the minority carriers (holes in n-type and electrons in p-type) can flow. They constitute the reverse or leakage current. This leakage is much higher in germanium than in silicon.

The width of the depletion region (Figure 3.5) is increased on application of the reverse voltage. This is because the free current carriers are repelled from the p-n junction.

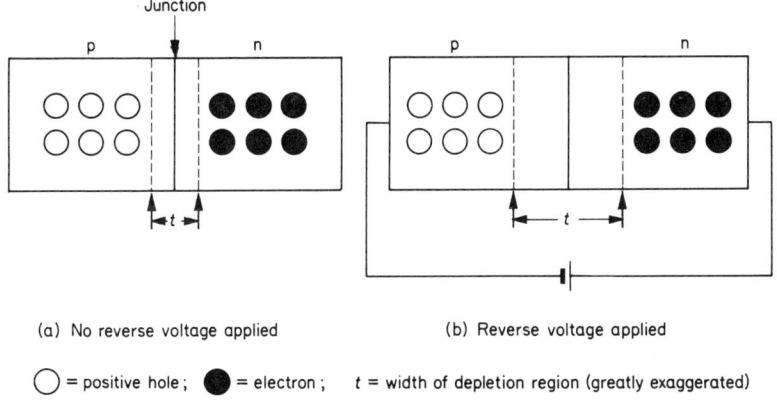

(a) No reverse voltage applied   (b) Reverse voltage applied

○ = positive hole ; ● = electron ; t = width of depletion region (greatly exaggerated)

Figure 3.5. The depletion region

The n region and the p-region separated by the depletion region form a parallel plate capacitor. The dielectric of this capacitor is the depletion region. The capacitance is increased if the area of cross-section of the p-n junction is increased and decreased if the depletion width is increased.

The fact that the capacitance of a junction diode is reduced by the application of a reverse voltage (depletion region width increased) has been exploited in the development of the *varactor diode*. Also, the capacitance of a reverse-biased junction is used to provide capacitance in an integrated circuit.

If a p-n junction diode is needed for operation at very high frequencies, the capacitance must be very small, otherwise the capacitive reactance X_c ($X_c = 1/(2\pi f C)$, where C is the capacitance and f is the frequency) will be small and significant unwanted high frequency alternating currents will flow. For detectors (essentially rectifiers) of very high frequency alternating potentials of small amplitude, point contact diodes of very small p-n junction cross-section area are used.

3.4 Brief Notes on Relevant Aspects of Alternating Current

To examine the performance of various rectifying arrangements it is necessary to recall the following facts:

(a) An alternating current of single frequency f and pulsatance $\omega = 2\pi f$, is represented by

$$i = I_p \sin \omega t \tag{3.3}$$

where i is the instantaneous current at time t and I_p is the peak current.

(b) The root-mean-square (r.m.s.) current I of an alternating current represented by equation (3.3) is given by

$$I = I_p/\sqrt{2}$$

I may be defined as the magnitude of the direct current which produces in a given time the same heating effect in a resistance R as the alternating current of peak value I_p. Similarly, for voltages,

$$V = V_p/\sqrt{2}$$

(c) Let V_p be the peak value obtained in a half-wave rectified voltage. The average value of this voltage is V_p/π. For a full-wave rectified voltage, the average value V_a is $2V_p/\pi$.

(d) For a diode, the *peak inverse voltage* (PIV)* is the maximum reverse voltage that it can withstand during the non-conducting interval. Exceeding the PIV will cause the diode to fail prematurely.

Note from (b) and (c) that for a full-wave rectified voltage of peak value V_p, the average value V_a is related to the r.m.s. value V by

$$V_a = 2V_p/\pi = 0.64V\sqrt{2} = 0.9V$$

* Now called the maximum repetitive peak voltage, for which the symbol is V_{RRM}.

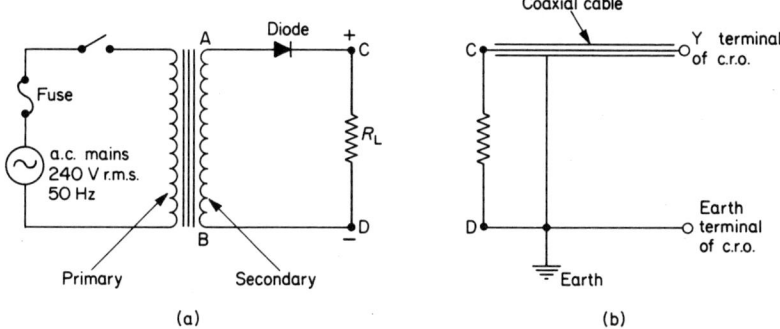

Figure 3.6. (a) Circuit diagram of a half-wave rectifier and (b) connection to a cathode ray oscillograph to demonstrate the waveform

3.5 Half-Wave Rectification

In the circuit used (Figure 3.6(a)) the primary winding of the mains transformer is connected across the a.c. mains (240 V r.m.s., 50 Hz). A fuse in series with this primary is a useful precaution. Across the secondary winding of this transformer is connected the load resistance R_L with the diode in series. Across R_L is produced a voltage which is unidirectional and consists of a series of pulses as shown in Figure 3.1(b).

With the diode connected as shown (the anode being joined to end A of the secondary winding and the cathode to end C of the load resistance R_L) it will conduct during each half-cycle when A is positive with respect to B, so there will be current through R_L. During the intervening half-cycles when A is negative with respect to B, the diode does not conduct (or rather conducts very poorly) so there is no or negligible current through R_L.

With a 50 Hz supply there will consequently be current pulses (each of half-cycle waveform) through R_L during, say, the first, third, fifth, seventh and so on hundredths of a second and corresponding voltage pulses developed across R_L. During the intervening second, fourth, sixth and so on hundredths of a second, the current through R_L will be zero for practical purposes and correspondingly no voltage will appear across R_L.

To demonstrate graphically the shape of the voltage pulses across R_L a cathode ray oscillograph (c.r.o.) is connected across R_L. The input to the Y-plates of the oscillograph is best made via a coaxial cable connection. The central wire of this coaxial cable connects end C of the load resistance R_L to the Y terminal of the c.r.o. The outer screen or metal sheath of the coaxial cable is connected to end D of R_L and to the earth terminal of the c.r.o. (Figure 3.6(b)).

Note that any point of the secondary circuit can be earthed as required because this secondary is isolated from the mains by the transformer.

Example 3.5

A half-wave rectifier circuit of the type shown in Figure 3.6(a) operates on an a.c. mains supply of 240 V r.m.s. with a step-down transformer of which the turns ratio T *is* $\frac{1}{3}$. *Calculate* (i) *the peak inverse voltage across the diode and* (ii) *the average value of the output voltage across the load resistance.*

The r.m.s. voltage across the secondary winding is

$$240 \times T = 240/3 = 80.$$

The peak secondary voltage $= 80\sqrt{2} = 113$. This will be the peak inverse voltage impressed across the diode during the non-conducting half-cycles. This is because the current through the load resistance is zero at these times and there is consequently no potential drop across this resistance. Hence the diode, when non-conducting, has to withstand the whole peak secondary voltage of 113 V.

The average output p.d. across the load resistance is given by

$$\text{(peak secondary voltage)}/\pi = 113/\pi = 36 \text{ V}$$

It is here assumed that the resistance of the conducting diode is very small compared with the load resistance so that, when it is conducting, the potential drop across the diode is negligible.

3.6 Full-Wave Rectification

The circuit (Figure 3.7) makes use of a mains transformer with a centre-tapped secondary winding, i.e. a terminal connection to the half-way point C in the winding is available so that the number of

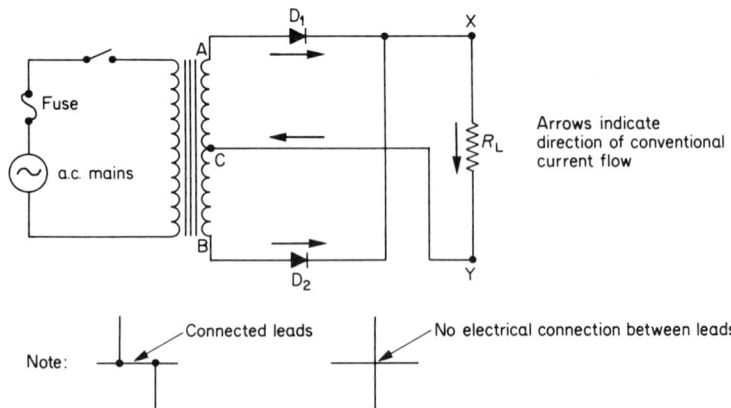

Figure 3.7. A full-wave rectifier unit

turns in the half AC equals that in half CB. End A of this secondary is connected via diode D_1 to one end X of the load resistance R_L. The other end B of the secondary is connected via a second similar diode D_2 *to the same point X.* The other end Y of the load resistance R_L is joined to the centre-tap C. Frequently, this point C is joined to earth.

The use of a centre-tapped secondary winding of a transformer is a simple and much used general method of obtaining two alternating voltages in anti-phase (exactly 180° out of phase). The potential at the centre-tap C is the reference point: this may or may not be earthed. Assume it is earthed for convenience so we may regard the potential of C as zero. When the alternating p.d. occurs across the whole secondary winding AB, at a particular instant of time suppose A is positive with respect to B. A will then be positive with respect to C, but B will be equally negative with respect to C. For example, if the instantaneous p.d. across AB is 100 V, that of A with respect to C will be $+50$ V, whereas that of B relative to C will be -50 V. This equal and opposite division will be maintained throughout: the potential of B with respect to C thus varies in anti-phase with that of A relative to C.

In the circuit of Figure 3.7, during a half-cycle when A is positive with respect to C, the diode D_1 conducts but as meanwhile B is negative with respect to C, diode D_2 does not conduct. In the immediately succeeding half-cycle, A will become negative relative to C whereas B becomes positive relative to C. Now the diode D_1 does not conduct whereas diode D_2 does. During both half-cycles note that the direction of the current through R_L is the same.

The unidirectional voltage across R_L thus varies with time as shown in Figure 3.1(c).

The average voltage V_a is $2V_p/\pi$, where V_p is the peak voltage across each half of the secondary winding.

The peak voltage across the whole secondary winding is $2V_p$. This will clearly be the peak inverse voltage across each diode.

If both diodes D_1 and D_2 were reversed, i.e. cathode of D_1 joined to A instead of anode and the same for D_2 in relation to B, the full-wave rectifier would still function. Now the centre-tap C would be the positive terminal in the conducting section and current would flow in the opposite direction through the load resistance.

3.7 The Full-Wave Rectifier Bridge Circuit

An alternative circuit to that of Figure 3.7 is one which dispenses with the centre-tap to the secondary winding and, instead, utilizes four diodes D_1, D_2, D_3 and D_4 in a bridge arrangement (Figure 3.8(a)).

In remembering how to draw this circuit note that the diodes in the four arms of the 'bridge' have their arrow heads all pointing 'upwards' (or they could all be 'downwards').

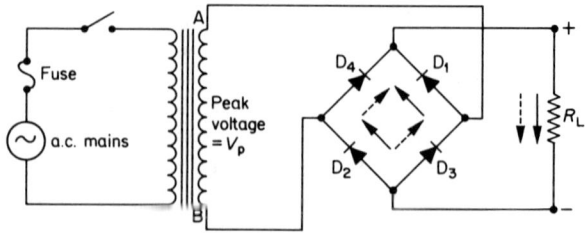

Full arrows indicate current flow when A+ w.r.t. B;
dotted arrows, the current flow when A − w.r.t. B.

(a)

(b)

(c)

Figure 3.8. A full-wave rectifier bridge circuit

During those half-cycles of the sinusoidal voltage variation across the transformer secondary winding when A is positive with respect to B, diodes D_1 and D_2 conduct (but D_3 and D_4 do not). The appropriate part of the circuit clarifying this statement is drawn in Figure 3.8(b). During the intervening half-cycles when A is negative with respect to B, diodes D_3 and D_4 conduct (but D_1 and D_2 do not).

The voltage waveform which appears across the load resistance R_L (Figure 3.8(c)) has each half-cycle labelled with the appropriate two diodes which are forward biased. In Figure 3.8(a), the full arrows indicate the direction of the current flow during those half-cycles when A is positive with respect to B and the dotted arrows are for the intervening half-cycles when the polarity of A relative to B is reversed.

As with any full-wave rectifier, the average output voltage across R_L is $V_a = 2V_p/\pi$. As V_p in Figure 3.8(a) is the peak voltage across the full secondary winding and this is applied across two diodes in parallel, when these diodes are non-conducting, the peak inverse voltage across each is V_p.

Certain features of bridge rectifiers are worth noting:

(i) Four diodes are required instead of the two of a centre-tapped transformer secondary circuit, but the average output voltage is $2V_p/\pi$ where V_p is the peak voltage across the whole secondary winding and not only half of it.

(ii) With two diodes in series with R_L during conduction, the potential drop across them is twice that for one only. As the potential drop across a semiconductor rectifier when forward biased is negligible, the use of two rectifiers in series is generally of no consequence in this connection.

(iii) If any one diode is faulty so that it conducts when reverse biased, a second diode is always damaged because of the excessive current flow generated. It is good practice to check the performance of the individual diodes before the bridge rectifier is constructed.

(iv) The bridge rectifier is compact and often used in rectifier-type moving-coil instruments for a.c. measurements.

(v) Most manufacturers of semiconductor components market bridge rectifiers as compact units. The four diodes are mounted and inter-connected appropriately on the necessary heat sinks and tested. The bridge rectifier is therefore quickly connected as a unit component with two input terminals and two output terminals.

3.8 A Voltage Doubling Circuit

On occasions when a high d.c. voltage is required and the current demands are moderate, a voltage doubler circuit may be used. Whereas with a full-wave bridge rectifier circuit the peak output voltage is V_p, a voltage doubler circuit may be constructed which provides a quite steady, output voltage of $2V_p$.

In the voltage doubling circuit (Figure 3.9(a)) two equal diodes D_1 and D_2 are connected in the same direction in series. Across them are connected two equal capacitances C_1^* and C_2 in series and also the

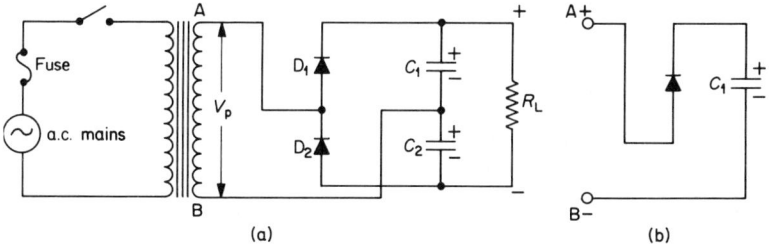

Figure 3.9. A voltage doubling circuit

---

* The circuit symbol for a capacitor has much thicker lines parallel to one another than in Figure 3.9 but thin lines are more easily drawn, *see also* Figure 4.13.

load resistance R_L. The transformer secondary winding is connected across the junction between D_1 and D_2 and the junction between C_1 and C_2.

If the load resistance R_L is large, corresponding to a small current drain on the circuit, the capacitances C_1 and C_2 in series will discharge through R_L at a much smaller rate than they are charged by the rectifiers D_1 and D_2.

During those half-cycles of the alternating potential difference across the transformer secondary AB when A is positive with respect to B, current flows through diode D_1 (but not through diode D_2) to charge up capacitance C_1. This is illustrated by part of the circuit drawn in Figure 3.9(b) in which the effect of the resistance R_L is omitted on the basis that the discharge current through it is comparatively very small.

A series of unidirectional pulses of current fed into a capacitor of capacitance C will, in time, provide a charge $Q = CV_p$, where V_p is the peak voltage obtaining at each pulse, presuming that the capacitor is not being discharged. This is rather like saying that a series of drops of water must eventually fill up a container if there is no hole in the container through which water can leak. If the current pulses are large enough and occur in rapid sequence additively, the capacitor voltage of V_p is quickly obtained.

During the intervening half-cycle when A is negative with respect to B, current flows through diode D_2 (but not through D_1), to charge up capacitance C_2.

Each of the capacitances C_1 and C_2 is therefore charged up to the peak voltage V_p across the transformer secondary. The polarity of the voltage across C_1 is the same as that across C_2. The p.d. set up across C_1 and C_2 in series, which equals that across the large resistance R_L, is therefore $2V_p$.

In practice there must be some discharge current through R_L. If this current is very small, the p.d. across R_L is $2V_p$ and very nearly constant. As R_L is made smaller, so the current through it increases, the voltage across it will drop and then rise again, so there will be fluctuations of this voltage. The extent of these fluctuations is clearly a question of balancing the rate of charge of C_1 and C_2 from the rectifiers against the rate of discharge through R_L.

3.9 An Alternative Voltage Doubling Circuit

In an alternative circuit to that of Figure 3.9, a capacitance C_1 is connected in series with the transformer secondary and a diode D_1, then across D_1 is connected a second diode D_2 in series with a second capacitor C_2. The load resistance R_L is across the capacitance C_2 (Figure 3.10).

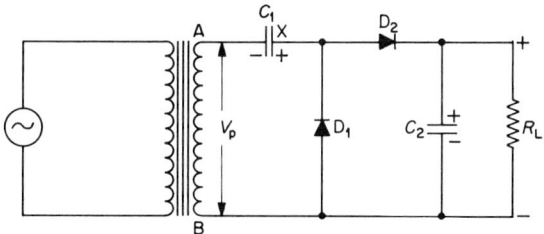

Figure 3.10. Voltage doubler alternative to that of Figure 3.9

During those half-cycles when B is positive with respect to A, diode D_1 conducts (but D_2 does not) so that capacitor C_1 is charged to the peak transformer secondary voltage V_p. Note that plate X of capacitance C_1 becomes positive with respect to A.

During the intervening half-cycles when A is positive with respect to B, diode D_2 conducts (but D_1 does not). The p.d. across D_2 and C_2 in series is now that across AB plus that across C_1: it is therefore $2V_p$. Hence C_2 becomes charged to a p.d. of $2V_p$.

As in the circuit of Figure 3.9(a), this action requires that the current drain (rate of discharge) through R_L is small compared with the rate of charging of C_1 and C_2 so that these capacitors can act as reservoirs of charge.

It is readily appreciated that in both these voltage doubler circuits (Figures 3.9 and 3.10) the peak inverse across each diode is $2V_p$.

3.10 A Voltage Quadrupler Circuit

Two circuits of the voltage doubler type shown in Figure 3.10 may be arranged as in Figure 3.11 to form a quadrupler circuit. Provided that

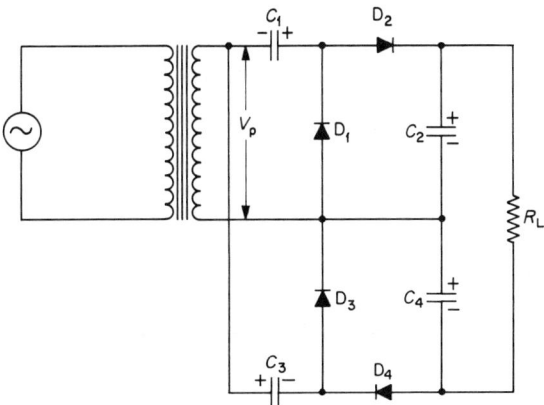

Figure 3.11. A voltage quadrupler circuit

the load current is very small (R_L very large) a p.d. across the load of $4V_p$ can be obtained.

3.11 Smoothing Circuits

After rectification (sections 3.5 and 3.6) the output voltage, although unidirectional, consists of pulses. Although these voltage pulses are approximately half of a sinusoidal wave form, the fluctuations which occur about the average voltage V_a are not sinusoidal. Nevertheless, the fluctuations repeat themselves exactly at regular intervals of time. The mathematical method of Fourier analysis (beyond the scope of this text) enables periodic waveforms of most shapes to be expressed in terms of sine waves of a fundamental frequency (first harmonic) and higher harmonics (second, third, fourth etc. of 2 times, 3 times, 4 times etc. respectively the fundamental frequency).

The rectified sine wave can thereby be shown to be represented by the summation of a number of simple components which include:

(i)   A steady voltage (or current, if current is the concern) of value equal to V_a, the average value. This average is V_p/π for half-wave and $2V_p/\pi$ for full-wave rectification (section 3.4).

(ii) An alternating voltage of which the frequency f is the same as that of the original alternating supply (at 50 Hz for the a.c. mains) which is rectified. This frequency f is the fundamental or the first harmonic.

(iii) Alternating voltages of frequencies $2f$, $3f$, $4f$ etc. which are correspondingly the second, third, fourth harmonics etc.

In an introductory account of smoothing circuits it is adequate to ignore the second and higher harmonic components of frequencies nf where n is an integer of two or more. Indeed, in practice, if the smoothing method is able to deal with the fundamental frequency component, it is generally more readily able to cope with the higher harmonic components.

What is required of a smoothing circuit, therefore, is a means of reducing significantly the alternating voltage of frequency f in the rectified output. This is achieved by a filter of which the simplest ones depend on a capacitor and a resistor. The filter has to be able to pass readily the d.c. component of the rectifier output and block the alternating component.

3.12 The Simple Capacitor-Resistor Smoothing Filter

A rectifier circuit such as is described in sections 3.5, 3.6 and 3.7 delivers across a load resistance R_L either a half-wave or full-wave rectified output, depending on its design. Basing the discussion on the half-wave rectifier, the circuit of Figure 3.6(a) with the waveform shown in Figure 3.1(b) is typical.

The object of the smoothing filter is to render steady the pulsating voltage across R_L. To achieve this with a simple capacitor filter, a capacitance C is connected across R_L and a safety resistance R_s is inserted in series between the diode and the capacitor to limit current surges.

The circuit is then as represented in Figure 3.12(a); the alternating input voltage has a peak value of V_p, that across the transformer secondary (the transformer is omitted in this diagram).

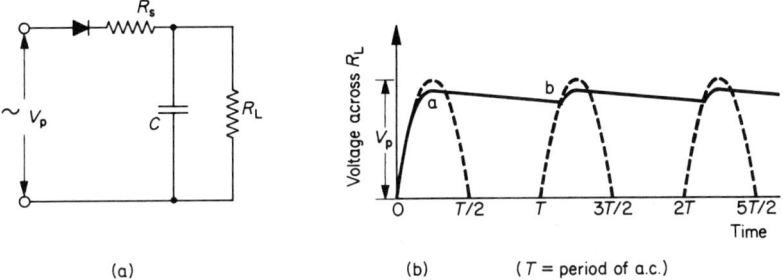

(a)   (b)   (T = period of a.c.)

Figure 3.12. A simple capacitor filter used with a half-wave rectifier

The capacitance C is charged by the rectified current pulses from the diode; it discharges through R_L. The rate at which it charges depends on the size of the current pulses passed through the diode. (It is assumed that the transformer used is capable of providing current of the magnitude needed). The rate at which C discharges decreases as the time constant CR_L increases. It is clearly important that a steady charge is maintained within C, which acts as a reservoir capacitor. To avoid too rapid a discharge rate of C, either C or/and R_L must be big.

Initially, the capacitance C is uncharged. When the circuit is first switched on, C may well charge up so rapidly that the current pulses through the diode are so large as to destroy it. This is avoided by the use of the safety resistor R_s, which limits the magnitude of charging current pulses*.

Suppose that CR_L is large. This capacitance C will then charge up almost to the peak voltage V_p during the first conducting half-cycle (Figure 3.12(b)).

During the immediately succeeding half-cycle, the charging current is zero (the diode is reverse biased). The voltage V_p across C will

---

* In power packs providing currents of up to 1 A, R_S is often omitted because a silicon rectifier can withstand a current surge of perhaps 20 times its average current rating.

therefore decrease exponentially because of the discharge through R_L. If CR_L is large enough, the decrease of voltage across C and R_L in parallel will be a small fraction of V_p during one period. Consequently the output voltage across C and R_L falls by only a small amount along ab and then, at b, it is restored almost to V_p again by the next charging half-cycle.

The voltage fluctuations (known as the *ripple voltage*) across C and R_L are therefore reduced to a small fraction of the magnitude of V_p that prevailed without the capacitance C. The percentage ripple is smaller the larger is the time constant CR_L.

For a full-wave rectifier, the discharge time is only half that for the half-wave so, other factors being the same, the percentage ripple is halved (Figure 3.13), and the ripple frequency is doubled.

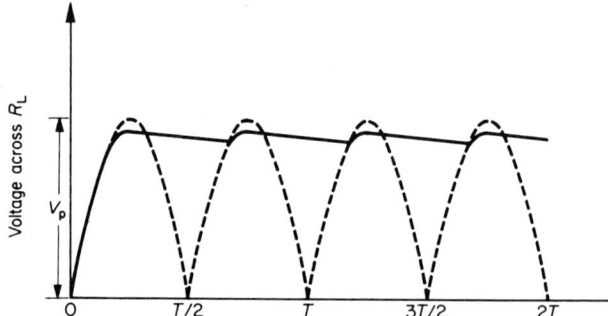

Figure 3.13. Full-wave rectified waveform showing the effect of the use of a smoothing capacitor

Another way of looking at the action of the smoothing capacitor is to consider that R_s ($\ll R_L$) and C are in series across the rectified a.c. supply which is equivalent to a steady voltage and an alternating voltage of frequency f (section 3.11). The reactance $X_C = 1/(2\pi f C)$ of the capacitance C is much less than the resistance R_L if C is large enough. The steady current output from the rectifier cannot pass through C so establishes a steady voltage across R_L. The alternating current component takes the much easier path through C so the fraction of this a.c. through R_L is much the smaller. Consequently, the voltage across R_L is the steady component subjected to a small alternating voltage: the ripple voltage.

Example 3.12(a)

A capacitor of capacitance 1000 μF is used to smooth the output from a half-wave rectifier. The transformer secondary output is 40 V r.m.s., 50 Hz, and the average rectified current (load current) is 15 mA. Calculate the average output voltage

and the peak value of the ripple voltage. Draw a graph showing the way in which the output voltage varies with time.

The peak voltage across the secondary winding is $\sqrt{2} \times 40 = 56$ V. The capacitor will charge to a maximum of very nearly 56 V during the conducting half-cycles and will discharge over a time of one cycle, which is 0.02 s.

The current through the load resistance $= 0.015$ A so that during 0.02 s, the quantity of electricity discharged is $0.02 \times 0.015 = 3 \times 10^{-4}$ C. The loss of charge of 3×10^{-4} C by the 1000 μF capacitor must correspond to a fall of the p.d. across it given by

$$\frac{3 \times 10^{-4}}{1000 \times 10^{-6}} = 0.3 \text{ V}$$

so the minimum voltage across the capacitor is $56 - 0.3 = 55.7$ V.

The average voltage across the capacitor, or average output voltage, is therefore $(56 + 55.7)/2 = 55.85$ V. The peak ripple voltage is half the fall of the capacitor voltage during one cycle and is hence 0.15 V. These results are exaggerated in the graph of Figure 3.14 assuming linear and not exponential voltage changes over the modest values encountered.

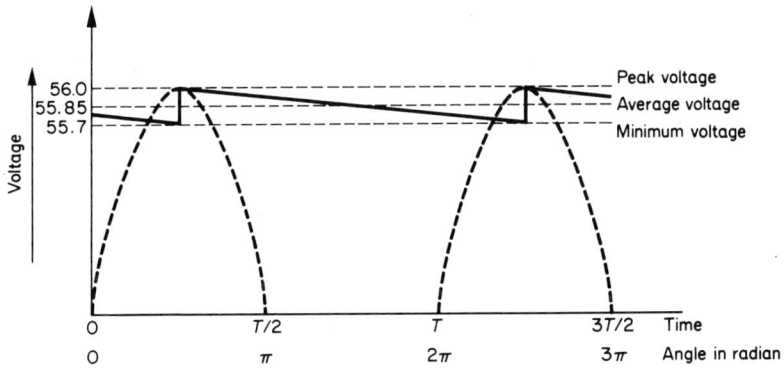

Figure 3.14.

Before the circuit of Example 3.12(a) is switched on the 1000 μF capacitor is uncharged. When switched on, this capacitor is charged to nearly 56 V when the charge it holds will be

$$56 \times 1000 \times 10^{-6} = 56 \times 10^{-3} \text{C}$$

This charge could be established in the first five milliseconds if the transformer winding impedance were very low. The instantaneous current through the diode would then be

$$56 \times 10^{-2}/5 \times 10^{-3} = 11.2 \text{ A}$$

This is an excessive current for a diode only required to pass 15 mA in continuous operation. Though 5 ms is perhaps an underestimate of the time, this calculation nevertheless illustrates the importance of two factors. One is that the surge current rating of a diode is an important characteristic to be noted. The second is that a safety resistance R_s is often a necessity to prevent damage when the circuit is first switched on.

Example 3.12(b)

A full-wave rectifier operating from a 50 Hz supply provides a peak output of 33 V. A load resistance of 10 kΩ is put across this output with a reservoir capacitor in parallel. Calculate the capacitance of this capacitor required if the ripple voltage is to have a peak value of 1 V.

The average voltage across the capacitor is 32 V. The average load current is (assuming linear discharge)

$$32/10^4 = 32 \times 10^{-4} \text{ A} = 3.2 \text{ mA}$$

The capacitor, on discharging for a time of half a cycle, which is 10^{-2} s, loses a charge Q of $3.2 \times 10^{-3} \times 10^{-2}$ C. The change of the p.d. across the capacitor during this time is to be 2 V for a peak ripple voltage of 1 V. Hence

$$2 = Q/C$$

where C is the capacitance required.

$$C = Q/2 = 3.2 \times 10^{-5}/2 = 16 \times 10^{-6} \text{ F}$$

i.e.
$$C = 16 \ \mu\text{F}.$$

3.13 Observations on Simple Capacitor-Resistor Smoothing

This method is simple and offers the advantages of a high output voltage and satisfactory smoothing if the load current is small (i.e. R_L is large). The disadvantages are:

(i) the ripple voltage increases with the load current;
(ii) the output voltage drops significantly as the load current increases, so the voltage regulation is poor.

Nevertheless, most small power packs use silicon p-n junction diodes and simple capacitor-resistor filters. To improve the voltage stability, a Zener diode (section 3.14) or a series control transistor (section 4.12) is employed.

3.14 Zener or Voltage Regulator Diodes

Zener diodes (also known as voltage regulator diodes or as reference voltage diodes) are silicon p-n junctions which are operated at a

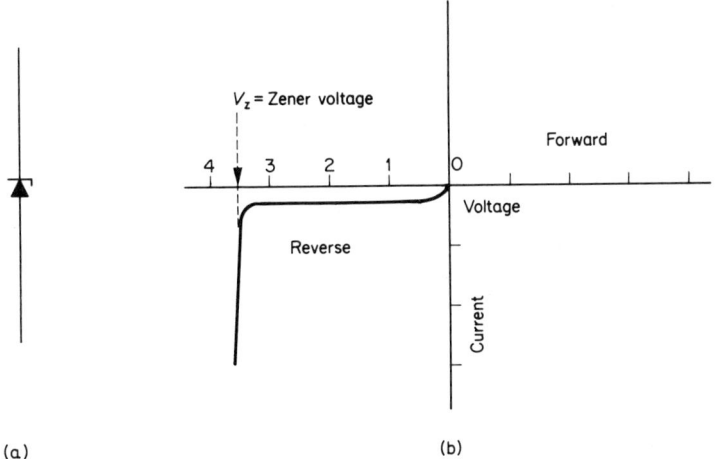

Figure 3.15. The Zener diode: (a) the circuit symbol; (b) the reverse bias
characteristic
(N.B. The reverse current is negligible until the Zener voltage is reached)

reverse bias value which is just beyond the junction breakdown
voltage. The circuit symbol is shown in Figure 3.15(a). To establish
reverse bias the external d.c. supply is connected so that the p-type
material (the 'anode') is negative and the n-type material (the
'cathode') is positive.

The reverse bias characteristic (Figure 3.15(b)) is like that of any
conventional p-n junction diode, as shown in Figure 3.3(d). The Zener
voltage V_Z is that at which the abrupt change in slope of the reverse
bias characteristic occurs. When the applied reverse bias voltage V is
less than V_Z, the resistance of the device is about 100 MΩ, and may be
assumed to be infinite for most practical purposes. When V is
increased to slightly beyond V_Z, the reverse current through the diode
increases very rapidly. The current is now limited only by the external
supply circuit. Hence in a circuit used to plot this characteristic and in
the use of the Zener diode to provide a reference voltage, a series
resistor must be included to prevent excessive current flow. Excess
current would cause overheating and failure of the junction.

When a reverse bias is applied across a p-n junction the width of the
depletion region is increased because free current carriers are repelled
from the junction (section 3.3). The width of the depletion region is
inversely proportional to the level of doping, which is equivalent to
saying that this width is proportional to the resistivity of the doped
semiconductor material.

If low-resistivity silicon (that with a high level of doping) is used, the
depletion region is very narrow. When a reverse bias is applied the

electric field strength across the depletion region about the p-n junction is therefore very large. If this reverse bias is increased to the Zener voltage V_Z, the electric field is so high that electrons are torn away from their normal locations about the atomic nuclei in the material. These electrons contribute massively to the current so that Zener breakdown results.

Zener diodes are thus p-n diodes with special depletion region characteristics, usually such that the Zener breakdown occurs below 5 V. The breakdown has to occur under controlled reversible conditions. Thus, if V exceeds V_Z breakdown occurs; when V is then reduced below V_Z, the normal high impedance p-n characteristic is restored, and this can occur repeatedly, indefinitely and reliably.

This breakdown mechanism is temperature dependent. On the application of a strong electric field, atoms release electrons to become ions more readily as the temperature is increased. Therefore Zener diodes with V_Z up to 5 V have a negative temperature coefficient: V_Z decreases somewhat as the temperature is increased.

With p-n junction diodes made from higher-resistivity doped silicon, the depletion layer is wider. Correspondingly, a higher p.d. across the depletion layer is needed to produce a given electric field strength. Now the minority carriers which constitute the reverse bias current can acquire sufficient energy to ionize lattice atoms on collision because they are accelerated through an adequate potential drop even though the electric field strength is not high enough to produce a Zener breakdown. The electrons produced by such ionization are themselves accelerated in turn and cause further ionization. This cumulative phenomenon results in a current avalanche.

There are consequently two different mechanisms: Zener breakdown proper due to an intense electric field: and ionization by collision resulting in a current avalanche. The former predominates with narrow depletion regions between low resistivity material for which V_Z is below 5 V. The latter — the avalanche breakdown — predominates with wider depletion regions between high resistivity material for which V_Z is above 5 V.

Nevertheless, voltage regulator diodes of both types are made and both are usually called Zener diodes, despite the fact that those with a breakdown voltage greater than 5 V do not operate primarily as a result of true Zener breakdown. In the depletion region, the mobilities of the current carriers decrease with increases of temperature. The avalanche breakdown phenomenon hence exhibits a positive temperature coefficient.

In general, therefore, Zener diodes with $V_Z < 5$ V have negative temperature coefficients and those with $V_Z > 5$ V have positive temperature coefficients. Zener diodes at which Zener voltages are between 5 and 6 can be selected to have effectively a zero temperature coefficient.

3.15 Zener Diodes in Voltage Regulator Circuits

A voltage regulator is a device which, when connected across a voltage supply source to a load, maintains constant the potential difference across the load despite fluctuations in the value of the load resistance or of the output voltage from the supply source.

A device is able to fulfil this purpose if it is such that, when the voltage across it exceeds some critical value, this voltage is independent of the current through the device.

The reverse bias characteristic of the Zener diode (Figure 3.15(b)) satisfies excellently this requirement. Once the Zener voltage V_Z has been reached, the characteristic is steep and linear. The slope of this linear portion is called the *slope resistance* of the diode.

Slope resistance

$$= \frac{\text{a small increase in the reverse bias voltage}}{\text{the corresponding increase in the current}}$$

$$= \Delta V / \Delta I$$

The lower the slope resistance, the more closely does the diode approach the ideal with a slope resistance of zero.

In choosing a voltage regulator diode, the manufacturer's data will provide:

(i) I_Z, the reverse current at which the characteristic was measured. This is obviously chosen so that the operating point on the characteristic (Figure 3.15(b)) is clear of the turnover point or 'knee'.

(ii) V_Z, the Zener voltage, e.g. minimum 5.3 V; maximum 5.9 V. There is always a 5 to 10 per cent tolerance in the manufacture of Zener diodes: they are not all exactly alike, but any one diode will always have a constant specific value of V_Z.

(iii) The typical temperature coefficient of the breakdown voltage expressed as per cent K^{-1}. In the voltage range from 5 to 6, these coefficients would be very small, i.e. < 0.001 per cent K^{-1}.

(iv) The slope resistance, e.g. 55 Ω.

(v) $I_{Z\,max}$ the maximum Zener current allowable at 40°C, e.g. 50 mA.

(vi) P_{max}, the maximum allowable power dissipation at 40°C, e.g. 300 mW.

Knowing this data and its current-voltage characteristic, it is necessary to decide the minimum current $I_{z\,min}$ to be clear of the 'knee'. Thus it might be decided that $I_{z\,min} = 1$ mA and that the Zener voltage $V_z = 5.8$ V.

3.16 A Simple Zener Voltage Stabilizer

In the circuit of Figure 3.16, V_i is the voltage from a power pack (e.g. a rectifier with capacitor smoothing), which may vary, and R_L is the

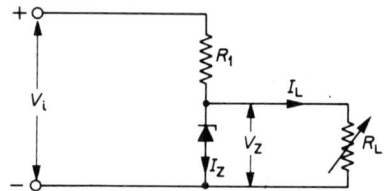

Figure 3.16. A simple Zener stabilizer circuit

load resistance (which may also vary) across which it is required to supply a constant voltage, irrespective of changes of V_i or of the load current I_L through R_L.

This is achieved by the use of a resistance R_1 in series with a Zener diode across the voltage supply V_i and the connection of R_L across the Zener diode. The value of R_1 is selected to determine the maximum current through the Zener diode with $I_L = 0$ and so R_L infinite.

Assume that the slope resistance of the Zener diode is zero, i.e. the diode is an ideal voltage regulator in that the appropriate part of the characteristic in Figure 3.15 is vertical when the reverse voltage across the diode is V_z. Used in a simple Zener stabilizer circuit (Figure 3.16) consider the circuit conditions when

(i) the input voltage remains constant but the load current changes;

(ii) the load resistance remains constant but the input voltage changes.

In both cases in Figure 3.16 suppose that, initially, $V_i = 18$ V, $V_z = 6.5$ V and $R_1 = 250$ Ω, then

the p.d. across $R_1 = 18 - 6.5 = 11.5$ V

the current through $R_1 = 11.5/250 = 46$ mA $= I_z$ for $I_L = 0$.

When a load resistance of $R_L = 1$ $k\Omega$ is connected across the output terminals

$$I_L = V_Z/10^3 = 6.5/10^3 = 6.5 \text{ mA},$$

so that

$$I_Z = 46 - 6.5 = 39.5 \text{ mA}.$$

Case (i). On decreasing the value of R_L the corresponding increase in I_L would cause an equal decrease in I_Z. This would continue until the Zener diode was no longer conducting and the circuit was no longer functioning as a voltage stabilizer.

Case (ii). On increasing the value of the input voltage V_i but keeping constant R_L, the corresponding increase in the current through R_1 would cause an increase in I_Z.

When I_L is a maximum, a minimum allowable current $I_{Z \text{ min}}$ must still flow through the Zener diode, and when I_L is zero the maximum current that flows through the Zener diode must not damage it.

The design considerations involved are illustrated by an example. Suppose a Zener diode having the typical characteristics given in section 3.15 is used. It is required to design a stabilizer which provides 5.8 V. It is assumed that the load current can vary between 0 and 10 mA and that the supply voltage V_i is nominally 9 V but never exceeds 10 V and must always be greater than 1 V above V_Z, i.e. > 6.8 V.

When the load current I_L falls to zero, the maximum current possible that flows through the Zener diode is $(I_{L \text{ max}} + I_{Z \text{ min}})$ where $I_{L \text{ max}}$ is the maximum load current encountered. This is 10 mA plus 1 mA (section 3.15), giving 11 mA.

Suppose the minimum p.d. across the series resistance R_1 is chosen to be 1.5 V. Then

$$R_1 = \frac{1.5}{11 \times 10^{-3}} = 136 \ \Omega$$

The nearest preferred value of R_1 selected from amongst the resistor values available from manufacturers is 150 Ω. Using this value, the maximum current $I_{Z \text{ max}}$ which flows through the Zener diode is when the input voltage V_i is at its maximum of 10 V and the load current I_L is zero. As $V_Z = 5.8$ V, so

$$I_{Z \text{ max}} = \frac{10 - 5.8}{150} = 0.028 \text{ A} = 28 \text{ mA}$$

This gives the maximum power dissipation in the Zener diode,

$$V_Z I_Z = 5.8 \times 0.028 \text{ W} = 162 \text{ mW}$$

which is well within the rated value of 300 mW.

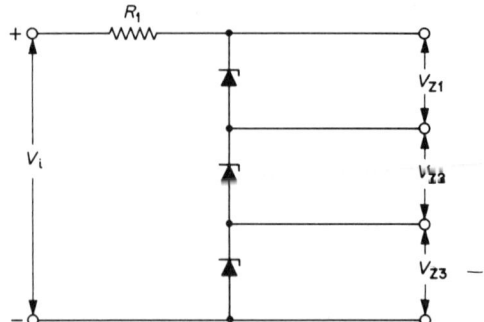

Figure 3.17. Three Zener diodes in series

3.17 Zener Diodes in Series and an Improved Stabilizer

The use of three Zener diodes in series (Figure 3.17) enables six different stable output voltages to be obtained. They are V_{Z1}, V_{Z2}, V_{Z3}, $(V_{Z1} + V_{Z2})$, $(V_{Z2} + V_{Z3})$ and $(V_{Z1} + V_{Z2} + V_{Z3})$. The current through any one of these series connected Zener diodes must not be allowed to fall below $I_{Z\,min}$.

The connection of Zener diodes as in Figure (3.18) enables very good stability to be achieved and is especially recommended if the supply voltage V_i contains an alternating ripple component. The

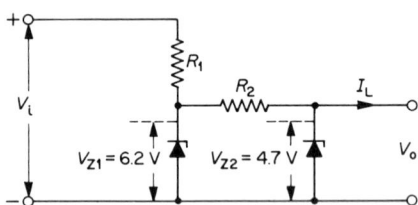

Figure 3.18 An improved voltage stabilizer using two Zener diodes

design is as in section 3.16 with each Zener diode being treated as a single stage. In a circuit such as that of Figure 3.18, it must, of course, be ensured that the Zener voltage for diode 1 exceeds that of diode 2, i.e. $V_{Z1} > V_{Z2}$ where V_{Z2} equals V_0.

3.18 A Variable Stabilized Supply

By making resistor R_2 of Figure 3.18 a variable component, two Zener diodes may be used in a circuit which provides a variable output which is stabilized between 4.7 V and 6.2 V, for example

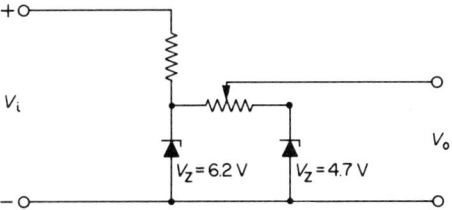

Figure 3.19 Two Zener diodes in a circuit to provide an adjustable stabilized voltage

(Figure 3.19). The stability is not so good as that obtained with the circuit of Figure 3.18.

3.19 Further Applications of Zener Diodes

Two more useful applications of Zener diodes are:

(i) To protect a voltmeter from excessive voltage: a Zener diode with V_Z at approximately twice that for full-scale deflection is conveniently used (Figure 3.20(a)).

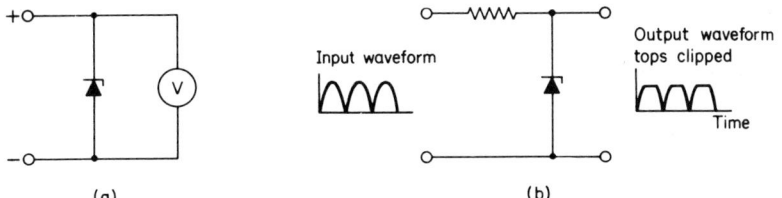

(a)   (b)

Figure 3.20 (a) Use of a Zener diode to protect a voltmeter from excessive overload and (b) a Zener diode limiter

(ii) As a means of limiting the height of a voltage pulse (Figure 3.20(b)). For example, the tops may be clipped from a unidirectional sinusoidal waveform voltage supply to produce a series of pulses of roughly rectangular waveform of constant peak voltage values.

3.20 Shunting a Current Meter to Provide a Non-Linear Scale

An interesting application of the silicon diode (*not* the Zener diode) is to use it in series with a resistance to shunt a microammeter. It is common practice to connect a low resistance in parallel with, say, a 0 → 100 μA meter to enable the meter to be used in a higher current range, say 0 → 10 mA. If this is done with the usual moving-coil meter,

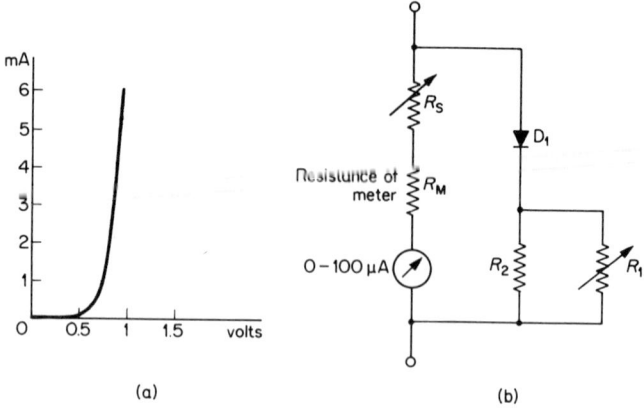

Figure 3.21 Arranging a current meter to have a non-linear scale

both the scales (with and without the shunt) are linear.

A silicon diode, of which the voltage-current characteristic is shown in Figure 3.21(a), in the shunt circuit (Figure 3.21(b)) enables the scale to be highly compressed over a certain current range whilst remaining approximately linear at each end, say between 0 and 15 μA at the lower end and between 1 and 10 mA at the high end. This provides a shunted meter which is convenient in many applications in that, on the one scale, a current can be recorded with an accuracy of about 5 per cent up to 15 μA and from 1 to 10 mA, and also gives an indication of the current between these extremes.

Initially a resistance R_s is connected in series with the resistance R_M of the meter to make $(R_M + R_s)$ equal to 12 kΩ approximately. In a suitable circuit with a current of 10 mA flowing, the variable resistance R_1 is adjusted to give a full-scale deflection on the 100 μA meter. With R_1 fixed at this value, the scale of the meter is calibrated against that of a good standard meter (e.g. an Avometer model 8). In general, the resistance R_M of the microammeter is approximately 1000 Ω so that the value of R_s required is approximately 11 kΩ. Any low-power silicon diode will serve for D_1 in Figure 3.21(b), whilst R_2 could be 82 Ω and R_1 a variable resistance of maximum value 1 kΩ.

The behaviour of the meter incorporating the non-linear shunt can be understood in a qualitative fashion from the diode characteristic (Figure 3.21(a)). When the voltage drop across the diode in the forward biased direction is small, say 0.1 V, the resistance of the diode is very high so negligible shunting of the meter occurs. The fact that the meter resistance including R_s is 12 kΩ in this small current range is of no significance because the total resistance in any circuit operating in this current range (say, 0→15 μA) is usually very large in comparison. As the voltage across the resistance of 12 kΩ $(R_s + R_M)$ increases, the diode current increases rapidly so the scale is very compressed. As the voltage across the diode increases, its effective resistance becomes very small and the 100 μA meter is shunted in the normal way.

When using a centre-zero meter to detect the balance condition in a Wheatstone bridge, it is usual to protect the meter movement by including a large series resistance which is short-circuited by a switch as the balance point is approached. Two silicon diodes can be used (to modify the meter movement in each direction) so that even an out-of-balance current of 10 mA would not damage the movement of a microammeter. In addition an out-of-balance current of 1 μA could still be detected, though with a slightly lower sensitivity than that produced by the microammeter alone.

Example 3.20

Referring to the circuit of Figure 3.21(b), calculate the effective resistance of the diode when a full-scale deflection of the 100 μA meter corresponds to a current of 10 mA, when $R_2 = 82$ Ω, $R_1 = 656$ Ω and $R_s + R_M = 12$ kΩ.

The appropriate circuit (Figure 3.22) includes the resistance R_D of the diode. The potential difference across $(R_s + R_M)$ is that due to 100 μA through 12 kΩ, i.e. it is $(12 \times 10^3 \times 10^{-4})\text{V} = 1.2$ V.

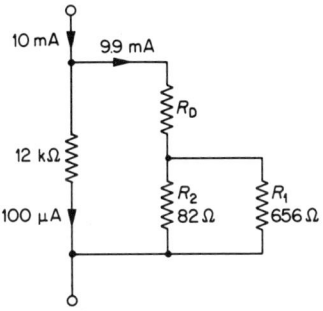

Figure 3.22

The current through the shunt must be 10 mA $-$ 100 μA $= 9.9$ mA. As the p.d. across the shunt must also be 1.2 V, it follows that the total resistance of the shunt must be

$$\frac{1.2}{9.9 \times 10^{-3}} = 120 \ \Omega$$

This must equal R_D in series with the combination of R_2 and R_1 in parallel. The resistance of this parallel combination is R given by

$$\frac{1}{R} = \frac{1}{82} + \frac{1}{656} = \frac{8+1}{656} = \frac{9}{656}$$

$$R = 656/9 = 73 \ \Omega$$

As $R_D + R = 120 \ \Omega$, hence R_D is 47 Ω.

Exercise 3

1. Explain the rectifying action of a p-n junction and compare the characteristic of such a junction with that of a vacuum diode.

(A.E.B., part.)

2. A forward-biased p-n junction diode at 300 K has a characteristic of which the voltage and current values are given in the table below:

Forward voltage (volt)	0.05	0.10	0.15	0.20	0.25	0.30
Forward current (ampere)	10^{-6}	5×10^{-6}	2×10^{-5}	10^{-4}	4×10^{-4}	1.5×10^{-3}

The relationship between the diode current I and the voltage V across the diode is given by the equation

$$I = I_0[\exp(eV/kT) - 1]$$

where e is the electronic charge, k is the Boltzmann constant, T is the absolute temperature and I_0 is a constant.

Present the values given in the table in the form of a graph suitable to verify this equation and calculate values for I_0 and for k. (Electronic charge $e = 1.6 \times 10^{-19}$C.)

3. It is required to identify the electrodes of an unmarked p-n junction diode and also to decide if it is made from germanium or silicon. Describe any experiment or tests necessary to achieve this.

(A.E.B., part.)

4. Draw circuit diagrams and explain the function of the components in a full-wave rectifier circuit using
   (a) a centre-tapped transformer;
   (b) a bridge-type rectifier.
   Outline the advantages and disadvantages of each.

5. In connection with a full-wave rectified waveform, define and state relations between the following values:
   (a) peak voltage;
   (b) root-mean-square voltage;
   (c) average voltage.

Draw the circuit diagram of a voltage doubling circuit and explain its operation.

6. A simple half-wave rectifier is connected to a sinusoidal a.c. supply of 100 V r.m.s. at 50 Hz, and is loaded with a resistance of 1000 Ω. Draw a diagram of the voltage waveform across the load resistance and calculate the peak value of the rectified current, assuming that the diode used is ideal.

7. A half-wave rectifier with a smoothing filter in the form of a capacitor of 8 μF is required to supply an average load current of 20 mA at an average voltage of 250. The mains supply is 240 V r.m.s. at 50 Hz. Calculate the transformer ratio required.

8. Explain the rectifying action of a silicon p-n junction diode. Draw circuit diagrams of full-wave rectifiers using (i) a centre-tapped transformer and (ii) a bridge-type circuit. Explain the action of a simple reservoir smoothing-capacitor.

   A full-wave rectifier of peak output voltage 25 V contains a simple reservoir smoothing capacitor of capacitance C. If the load on the rectifier is 4 kΩ calculate the value of C required to ensure that the peak ripple voltage does not exceed 1.0 V. Frequency = 50 Hz.

   (A.E.B.)

9. Explain the terms *depletion layer, minority carriers* and *reverse saturation current*. What factors determine the magnitude of the Zener voltage and the temperature coefficient of a reference voltage diode?

10. Distinguish between Zener and avalanche breakdown in voltage regulator diodes.

    Explain, with the aid of a circuit diagram, how a Zener diode can be used to provide a steady voltage across a load, despite the fact that both the input voltage and the load resistance can vary.

11. It is required to make the response non-linear of a microammeter which has a full-scale deflection of 100 μA. The object is to achieve a scale which is linear in the region 0–15 μA and also that the meter will indicate currents of up to 10 mA without the use of a range switch. Explain how this can be achieved.

4 Transistors and some basic applications

4.1 Bipolar Junction Transistor

The bipolar transistor — a very important control device in electronics — consists of two p-n junctions positioned close together within a single crystal slice. The region common to the two junctions, called the *base*, may be either of n-type semiconductor material or of p-type. In the first case, the n-type base region is flanked on each side with p-type material, forming a p-n-p junction transistor. In the second case, the p-type base region is flanked by n-type material, forming an n-p-n junction transistor. The thin central region — the base (b) — is of high resistivity material sandwiched between more heavily doped material comprising the *emitter* (e) on one side and the *collector* (c) on the other side. The most commonly encountered bipolar transistors are n-p-n ones in silicon.

The structures are illustrated in outline in Figure 4.1(a). The distinction between the conventional circuit symbols of the two types (Figure 4.1(b)) is made by the direction of the arrowhead on the line joining the emitter to the base. This arrowhead points in the direction of *conventional* current flow (i.e. from positive to negative, opposite to that of electrons). Thus, in the p-n-p transistor, the arrowhead is directed from emitter to base because positive charge flows readily from p to n across the junction, whereas in the n-p-n transistor, the arrowhead points from base to emitter.

The two junctions formed are the emitter-base junction (J_{eb}) and the collector-base junction (J_{cb}). Each of these junctions behaves in a similar way to the p-n junctions already described in Chapter 3. However, provided that the junctions are correctly biased (i.e. have small voltages of the correct polarity across them), additional and most useful behaviour results from these junctions being so close together, separated only by the thin base region.

To obtain transistor action the emitter-base junction must be forward biased and the collector-base junction reverse biased.

To obtain forward bias across the emitter-base junction of a n-p-n transistor, the base must be made positive with respective to the emitter; electrons then flow readily from the n-type emitter to the p-

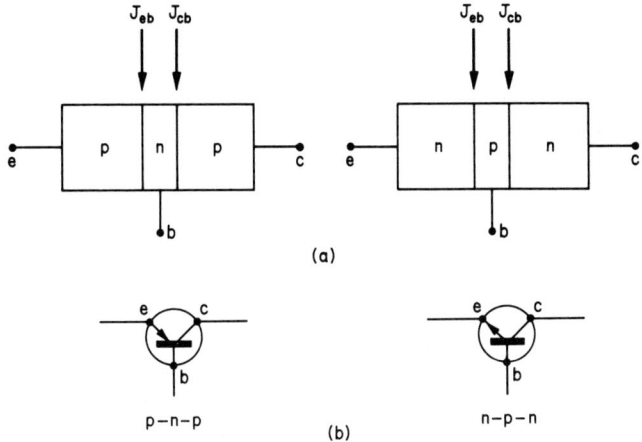

Figure 4.1 (a) Schematic diagrams of p-n-p and n-p-n junction transistors and (b) circuit symbols for these transistors

type base across the junction (J_{eb}). For a p-n-p type transistor, the base has to be negative with respect to the emitter to provide forward bias so that now holes flow readily from emitter to base.

The behaviour of the n-p-n transistor is considered in detail. That of the p-n-p type will be apparent, bearing in mind that the applied voltages will be of opposite polarity and the majority carriers will be of opposite charge.

On applying forward bias to the emitter-base junction of an n-p-n transistor, the consequent movement of electrons from the emitter into the p-type base region causes more electrons to move into or be injected into the emitter to maintain equilibrium. Once in the base region, the electrons do not have far to travel by diffusion through the thin base before they are accelerated across the collector-base junction (J_{cb}) to the positive collector electrode.

Some electrons will combine with holes in the base region and so are unable to reach the collector. Electrons which move into the base to compensate for this loss due to recombination will constitute the base current I_B.

The greater the reverse bias across the collector-base junction, the further will the depletion layer (section 3.3) extend into the p-type base. This effectively decreases the width of the base and reduces the number of electrons lost by recombination.

To provide current flow in a forward biased p-n junction in silicon a p.d. of about 0.7 V is needed (section 3.2).

A junction transistor may be connected in one of three ways: *common-base*, *common-emitter* or *common-collector*. As the last of

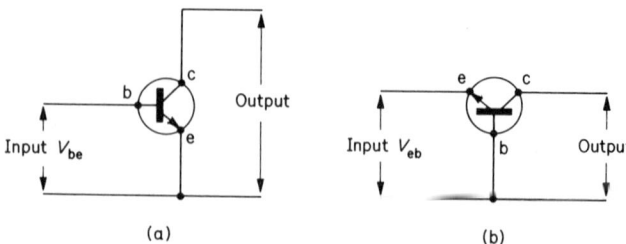

Figure 4.2. (a) Common-emitter connection and (b) common-base connection

these is the least used it is not considered further here. Common-emitter connection (Figure 4.2(a)) is the most widely used method; the common-base connection (Figure 4.2(b)) is not so versatile. The bias across base-emitter is V_{BE}, across collector-emitter it is V_{CE} and across collector-base, V_{CB}. The electrode stated to be common is in fact that one which is common to both the input and the output circuits.

4.2 The Characteristics of an n-p-n Transistor in Common-Emitter Connection

The circuit of Figure 4.3(a) enables the characteristics of n-p-n transistors in common-emitter connection to be plotted. An example is a BC107. The base connections are shown in Figure 4.3(b). *Voltage supplies should not be connected until the polarity has been checked against the circuit diagram.* The transistor could be permanently damaged if a junction bias were incorrect.

The voltmeters used to record V_{BE} and V_{CE} should have a high resistance of 10 MΩ.* If multi-range instruments are used for the current measurements (I_B and I_C), their range should not be altered during an experiment.

The meter used to record the base current I_B may have a full scale deflection (f.s.d.) of 250 μA, whilst that for the collector current I_C may have an f.s.d. of 25 mA.

The *input characteristics* (Figure 4.4) are those of the base-emitter voltage (input voltage) V_{BE} plotted against the base current (input current) I_B for a number of values of the collector-emitter voltage V_{CE}.

The *input resistance* of the transistor is not a constant except at given values of V_{CE} and I_B (or V_{BE}). At a chosen point on a given input characteristic, the input resistance† (h_{ie}) is given by

$$h_{ie} = \Delta V_{BE}/\Delta I_B$$

---

* Digital voltmeter (DVM) is ideal: see Appendix
† The term should strictly be the 'incremental resistance' or 'slope resistance' because the small *change* of current accompanying a small change of voltage is concerned. However, the adjective 'incremental' is not employed in common practice in this case.

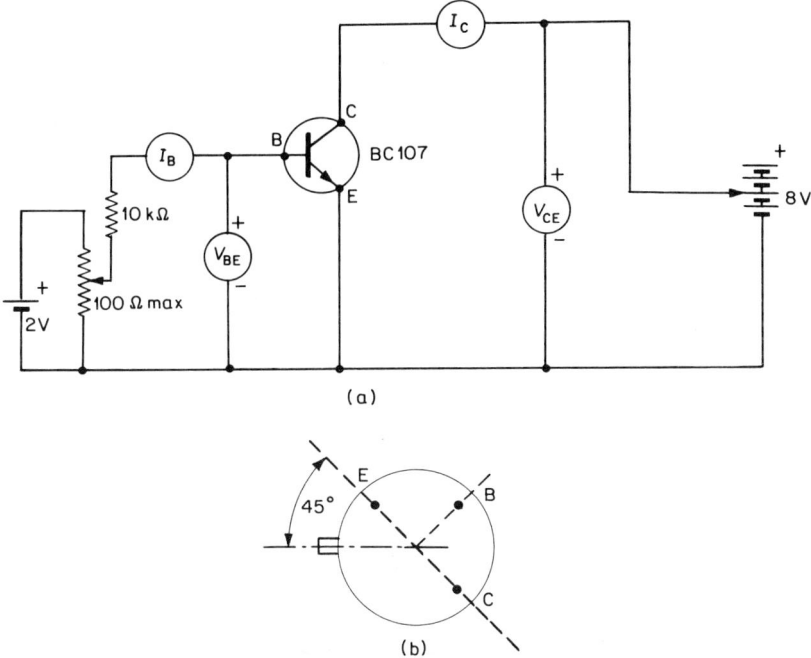

Figure 4.3. (a) Circuit for plotting the characteristics of an n-p-n transistor in common-emitter connection and (b) the electrodes of the BC107, viewed from the underside (a TO18 housing), the base electrode B is connected internally to the metal case

where ΔV_{BE} is a small change of V_{BE}, and likewise for ΔI_B; h_{ie} is therefore decided by the slope of the input characteristic at a chosen point. Since h_{ie} has the dimensions of resistance, it is quoted in ohm. For example, the line LM (Figure 4.4) is a tangent to the curve at the point for which $V_{CE} = +2.0$ V and $I_B = +50\ \mu A$ and could be chosen to evaluate h_{ie}.

In specifying the parameter h_{ie}, the subscript i denotes input and the subscript e signifies common-emitter connection. The value of h_{ie} may be between 1600 and 15000 Ω. It is approximately eight times the input resistance in common-base connection, which would be denoted by h_{ib}.

The *output characteristics* (Figure 4.5) of the transistor in common-emitter connection are of the collector (output) current I_C plotted against the collector-emitter (output) voltage V_{CE} for a number of values of the base current I_B.

The slope of the output characteristic, say at the reference point Q

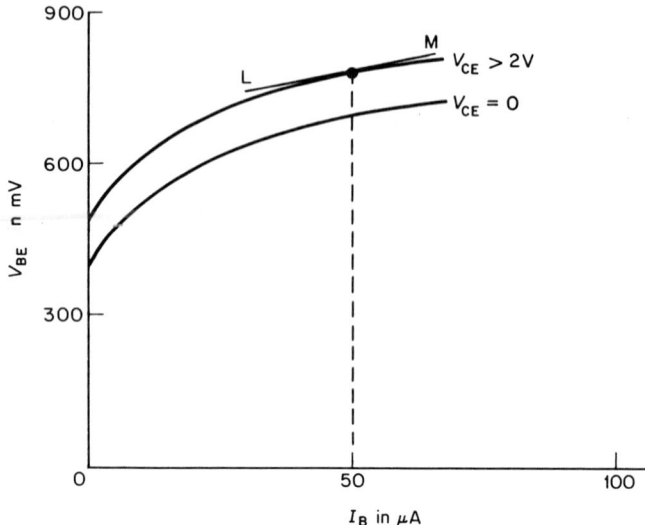

Figure 4.4. Typical common-emitter input characteristics for an n-p-n transistor

in Figure 4.5 is denoted by h_{oe}, where the subscript o denotes 'output'. Thus

$$h_{oe} = \Delta I_C / \Delta V_{CE} \quad \text{at constant } I_B *$$

This parameter has the dimensions of conductance; it is the output conductance of which the reciprocal is the *output resistance*. The output resistance of the transistor is very high, of the order of 30 kΩ.

The relatively low input resistance due to the forward bias across the input junction (J_{EB}) compared with the high output resistance due to the reverse bias across the output junction (J_{CB}) gives rise to the term 'transfer-resistor' or 'transistor'.

The most important parameter of the transistor is concerned with the current gain. Denoted by h_{fe} where the subscript *f* denotes 'forward', it is called the *forward current transfer ratio* (or current gain). It must be ensured in specifying this current gain that the collector-emitter voltage V_{CE} remains constant. In practice, therefore, it is asserted that the current gain is with the output short-circuited to alternating current.

* To denote a change of *V* or of *I*, the symbol Δ is used, as it is familiar. In conventions in electronics, the capital letter is used for the symbol if the voltage or current is steady, whereas the small (lower-case) letter is used for a changing voltage or current, usually alternating or a pulse. In this text if a capital letter is used for *V* or *I*, the quantity can be either steady or varying.

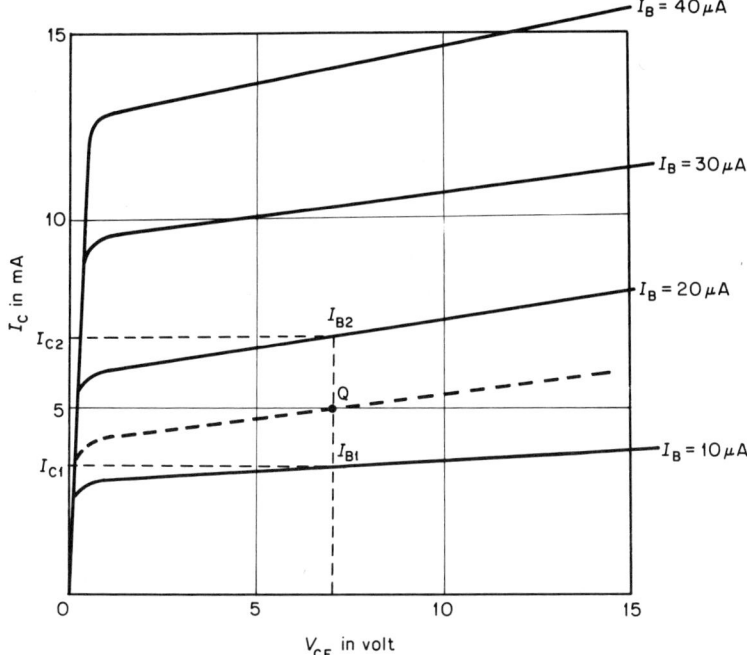

Figure 4.5. Typical output characteristics of an n-p-n transistor, BC107
[about point Q: $h_{fe} = (I_{C2} - I_{C1})/(I_{B2} - I_{B1}) = (6.9 - 3.4) \text{mA}/(20 - 10)\mu\text{A} = 350$]

$$h_{fe} = \Delta I_C / \Delta I_B$$

Referring to the linear portions of the characteristics of Figure 4.5, it is seen that

$$h_{fe} = \frac{(I_{C2} - I_{C1})}{(I_{B2} - I_{B1})} \quad \text{at a constant value of } V_{CE}$$

The *transfer characteristic* (Figure 4.6) is obtained by plotting the collector current I_C against the base current I_B for a given value of the collector-emitter voltage V_{CE}, say at $+2$ V.

The transfer characteristic is a straight line. Hence

$$I_C = \text{constant} \times I_B$$

This constant is denoted by β or h_{FE}, where

$$h_{FE} = I_C / I_B$$

Note the similarity and at the same time the important difference between h_{fe} and h_{FE}. The former is at a given operating point and is

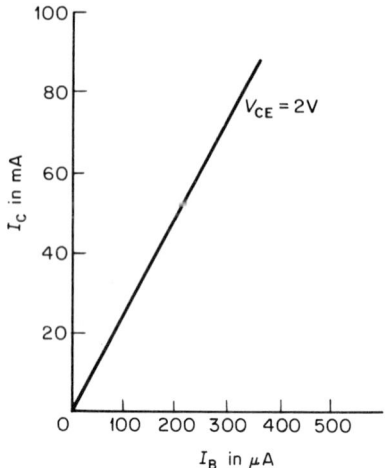

Figure 4.6. Typical transfer characteristic for an n-p-n silicon transistor in common-emitter connection

concerned with small changes of I_C and I_B. The latter —distinguished by the use of capital letter subscripts — is a constant obtained from a linear graph and is called the *static value of the forward current transfer ratio* or, more simply, the *direct current gain*. It is the ratio between the continuous output current (in this case I_C) and the continuous input current I_B.

In the experiment, the values of h_{fe} and h_{FE} should be compared. They are almost equal and in future in this text will be thought of as equal. For the BC107 they may have a value of about 250.

4.3 Characteristics of an n-p-n Transistor in Common-Base Connection

The circuit of Figure 4.7 enables the characteristics to be obtained of an n-p-n transistor in common-base connection. Again, the polarities

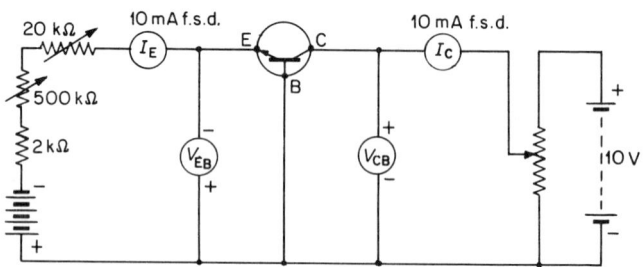

Figure 4.7. Circuit for obtaining the characteristics of an n-p-n transistor in common-base connection

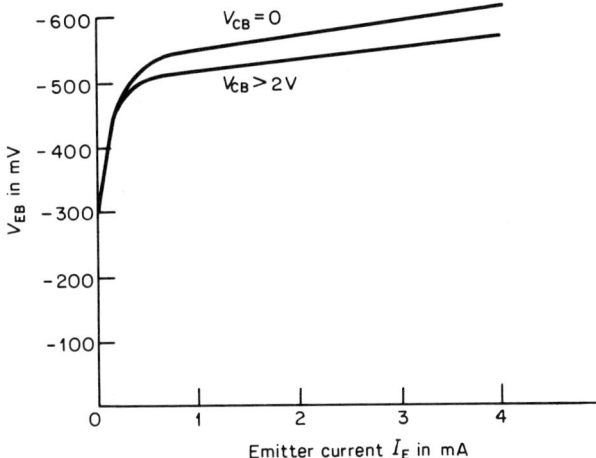

Figure 4.8. Typical input characteristics of an n-p-n transistor in common-base connection

of the applied voltages should be carefully checked before the circuit is connected.

The input characteristics (Figure 4.8) show the relationship between the emitter-base potential difference V_{EB} and the emitter current I_E for two collector-base voltages, V_{CB}.

The *input resistance*, denoted by h_{ib}, where the subscript i denotes 'input' and b denotes 'base', is not constant but will depend on the chosen value of the emitter current I_E and the collector-base voltage V_{CB}. It is given by

$$h_{ib} = \Delta V_{EB}/\Delta I_E \quad \text{at constant } V_{CB}$$

and is obtained from the slope of the tangent at the operating point on the curve. If ΔV_{EB} is in volt and ΔI_E in ampere, the input resistance h_{ib} is given in ohm.

The value of the h_{ib} is between 50 and 100 ohm. This low input resistance is a consequence of the forward biased emitter-base junction and is, in many circumstances, a disadvantage of junction transistors.

The output characteristics can be obtained by setting the emitter current (I_E) at some fixed value (say 1 mA) and measuring the collector current (I_C) for various collector-base voltages (V_{CB}) between, say, 0 and +9 V. The family of output curves (Figure 4.9) shows that the output current (I_C) varies only very slightly with the collector-base

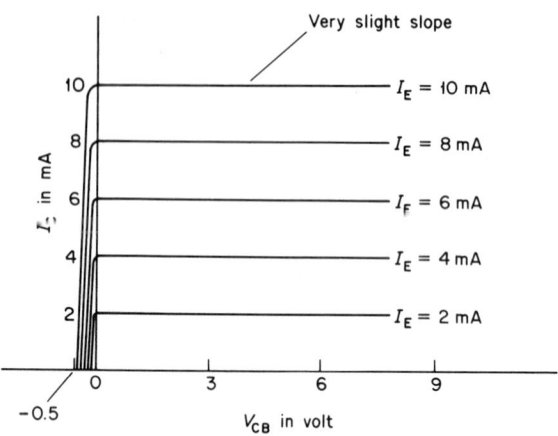

Figure 4.9. Typical output characteristics of an n-p-n transistor in common-base connection

voltage (V_{CB}), the characteristics being straight lines almost parallel to the V_{CB} axis. The slope of these lines gives the output conductance, denoted by h_{ob}, where the reciprocal is the output resistance of the transistor. Hence

$$h_{ob} = \Delta I_C / \Delta V_{CB} \quad \text{at constant emitter current.}$$

The output resistance is very high, certainly greater than 0.5 MΩ and is explained by the fact that the collector-base junction is reverse-biased.

In common-base connection the transistor acts as a constant current generator. This is because the output current (I_C) varies only slightly with the collector-base voltage (V_{CB}). Hence if a load resistance is inserted in series with the supply voltage and the collector, although this will reduce the voltage V_{CB} because of the p.d. produced across this resistance, the current through the resistance will remain almost constant even when the value of this resistance is altered.

The forward current transfer ratio, h_{fb}, also known as the forward current gain, for the transistor in common-base connection can be obtained — although not very accurately — from the curves of Figure 4.9.

$$h_{fb} = \frac{\Delta I_C}{\Delta I_E} = \frac{I_{C2} - I_{C1}}{I_{E2} - I_{E1}} \quad \text{at constant } V_{CB}$$

where ΔI_C is the change of the collector current I_C for a small change ΔI_E of the emitter current I_E. The value of h_{fb} is approximately 0.99.

4.4 Some Features of the *h* (hybrid) Parameters

The parameters of the transistor so far determined from the input and output characteristics are termed *hybrid* or *h* parameters because they are not all alike dimensionally. Thus, h_{ie} and h_{ib} have the dimensions of resistance, h_{oe} and h_{ob} have the dimensions of conductance and the remaining ones, h_{fe} and h_{fb} are dimensionless.

To recapitulate briefly, the subscripts employed denote the following: the first letters i, o and f denote respectively input, output and forward current; the second letters e and b denote respectively common-emitter and common-base connections.

These hybrid parameters are connected with small variations of the voltage or current about a specific operating point. They are of importance when small signal variations (the special case being small signal a.c. behaviour) are applied to the transistor. They are easy to measure and of value in enabling the performance of a transistor amplifier to be calculated. The correct terms involved are:

(a) In common-emitter connection,

$$h_{ie} = \left(\frac{\Delta V_{BE}}{\Delta I_B}\right)_{V_{CE}}$$

which is the input resistance with the output short-circuited to a.c. so that V_{CE} remains constant;

$$h_{oe} = \left(\frac{\Delta I_C}{\Delta V_{CE}}\right)_{I_B}$$

the output conductance with the input open-circuit to a.c. so that I_B remains constant;

$$h_{fe} = \left(\frac{\Delta I_C}{\Delta I_B}\right)_{V_{CE}}$$

the forward current transfer ratio (or current gain) with the output short-circuited to a.c. so that V_{CE} remains constant.

(b) For common-base connection,

$$h_{ib} = \left(\frac{\Delta V_{EB}}{\Delta I_E}\right)_{V_{CB}} \quad h_{ob} = \left(\frac{\Delta I_C}{\Delta V_{CB}}\right)_{I_E} \quad h_{fb} = \left(\frac{\Delta I_C}{\Delta I_E}\right)_{V_{CB}}$$

The d.c. biasing of a transistor is considered later (section 4.8). These hybrid parameters are not concerned with the biasing; however, they *are* concerned with small signal variations about the operating points determined by the bias. For such small signal variations, linear behaviour of the transistor about the operating point is assumed.

Manufacturers usually publish the characteristic curves of their junction transistors and quote values of h_{fe} at a specific value of I_C, say 1 mA. In general, between transistors of the same type, there is a large spread in the values of a particular parameter.

The output resistances in both CE and CB connections ($1/h_{oe}$ and $1/h_{ob}$ respectively) are very high. In the simple basic amplifier circuit an output load resistance R_1 is connected in series between the voltage supply and the collector. The object is to obtain a voltage variation (signal) across this load which is an amplified replica of the input signal. This load resistance between collector and the battery supply terminal is in parallel with the output resistance of the transistor as regards any signal variation. This is because the internal resistance of the battery supply is negligibly small. Hence, in any analysis, the output resistance of the transistor may be considered to be infinite provided that the load resistance R_L is not too high. In practice this means that R_L should not be more than 10 per cent of $1/h_{oe}$ or $1/h_{ob}$.

The transistor is then behaving as a constant current generator working into a load resistance R_L.

4.5 Amplification in Common-Emitter Connection

To show how a junction transistor in common-emitter connection is able to amplify a small voltage change applied at its input, consider Figure 4.10. That this circuit can amplify is due to the fact that a small

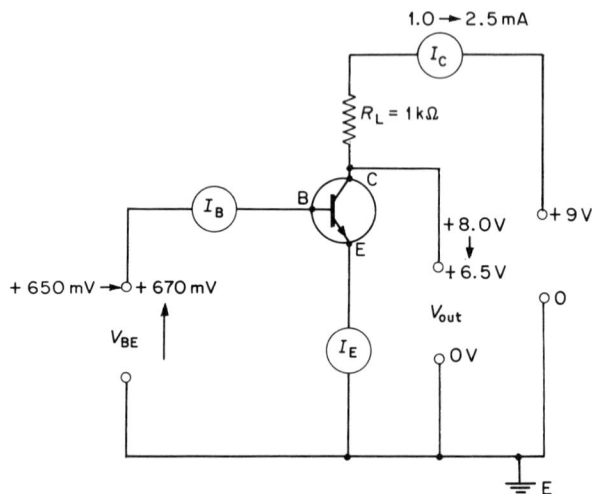

Figure 4.10. Behaviour of a voltage amplifier based upon an n-p-n transistor in common-emitter connection

change ΔI_B of the base current I_B can produce a large change ΔI_C in the collector current I_C, as shown in section 4.2, because h_{fe} is about 250.

The emitter — which is common to both the input and the output circuits — is considered to be at zero potential: it is often earthed, although this is by no means essential. As the transistor is of the n-p-n type, its collector potential must be positive with respect to earth. The supply voltage to the collector is therefore from a battery supplying a maximum of $+9$ V. However, it is of no use in designing an amplifier to join the collector directly to the supply voltage because it is essential for the collector potential to vary in sympathy with input voltage changes. Therefore, a load resistance R_L is connected in series between the $+9$ V terminal of the supply and the collector. It is across R_L that the amplified version of the input voltage change is to be produced. If R_L is 1 kΩ, it will have little effect on the collector current because the output resistance of the transistor is very high.

To avoid, for the time being, the consideration of bias supplies to the input, suppose the base-emitter voltage V_{BE} changes from $+650$ mV to $+670$ mV. As the input voltage becomes more positive, the base current increases. This increase might be from 5 μA to 12.5 μA if, say, a BC107 is used. As a result the collector current I_C increases from, say, 1.0 mA to 2.5 mA. When the collector current is 1.0 mA the p.d. across R_L of 1 kΩ is $1.0 \times 10^{-3} \times 10^3 = 1.0$ V. As the battery supply e.m.f. is $+9$ V, the collector potential will therefore be $+8.0$ V. When the input potential V_{BE} is altered to $+670$ mV (the change of V_{BE} is $+20$ mV), the collector current increases to 2.5 mA. The p.d. across R_L therefore increases to 2.5 V and so the collector potential becomes $+6.5$ V.

Consequently, a change of voltage of 1.5 V is produced across the load resistance R_L for a change of the input voltage by 20 mV. This corresponds to a voltage amplification ratio (voltage gain) of 1.5/0.02 $= 75$. Several features of this simple amplifier are of interest.

(a) If the input voltage is increased negatively, the collector voltage becomes more positive, and *vice versa*, if the input is increased positively, the collector voltage becomes more negative. Hence if the input voltage is alternating, the output voltage across the collector-emitter is 180° out of phase with the input voltage. The amplifier thus introduces a phase change of π or 180°.

(b) The input resistance of the amplifier is defined simply as the change in input voltage, ΔV_{BE}, divided by the resulting change in input current, i.e. in the present example,

$$\frac{\Delta V_{BE}}{\Delta I_B} = \frac{20\ mV}{7.5\ \mu A} = 2600\ \Omega$$

(c) The current gain, denoted by A_i, is defined by

$$\frac{change\ in\ collector\ current}{change\ in\ base\ current} = \frac{\Delta I_C}{\Delta I_B}$$

$$= 1.5\ mA/7.5\ \mu A = 200$$

(d) The voltage gain, denoted by A_v, is defined by

$$\frac{\Delta V_{out}}{\Delta V_{in}} = \frac{\Delta V_{CE}}{\Delta V_{BE}} = \frac{1.5\ V}{20\ mV} = 75$$

(e) The power gain is defined as

$$\frac{change\ in\ output\ power}{change\ in\ input\ power} = A_v \times A_i$$

which, in the present example is $75 \times 200 = 15000$.

4.6 A.C. Amplification and the Load-Line: A Graphical Analysis

When an input alternating or pulsating signal (voltage input) is to be amplified, the usual requirement is that the output from the amplifier be a true magnified replica of the input. This means that it must have the same waveform or pulse shape. A problem is that the input characteristics (I_B against V_{BE} with V_{CE} as parameter) are curved; the input resistance of the amplifier is not constant. Hence the input signal voltage must be kept small and take place about as linear a part of the input characteristic as possible. Thus a steady (so-called, d.c.) bias voltage is necessary to ensure this and the magnitude of this voltage is considerably larger than the peak value of the input signal. It is also important to ensure that the bias current (base current) is steady.

The required d.c. bias can be provided in various ways (section 4.8). Let us assume that this bias is made available, as represented in the circuit diagram of Figure 4.11(a), where the input signal is only 30 mV. The d.c. bias in this circuit is +650 mV.

Two points demand the introduction of the capacitors C_1 and C_2 in this circuit. The first is that the input signal source cannot be joined directly across the +650 mV bias supply. If it were (i.e. C_1 were omitted) the result would be simply the passage of direct current through the signal source — the result would be most unsatisfactory: to avoid this is the function of C_1. This has a reactance X_{C1} given by $1/2\pi f C_1$ at the frequency f of the input signal. At the same time, the

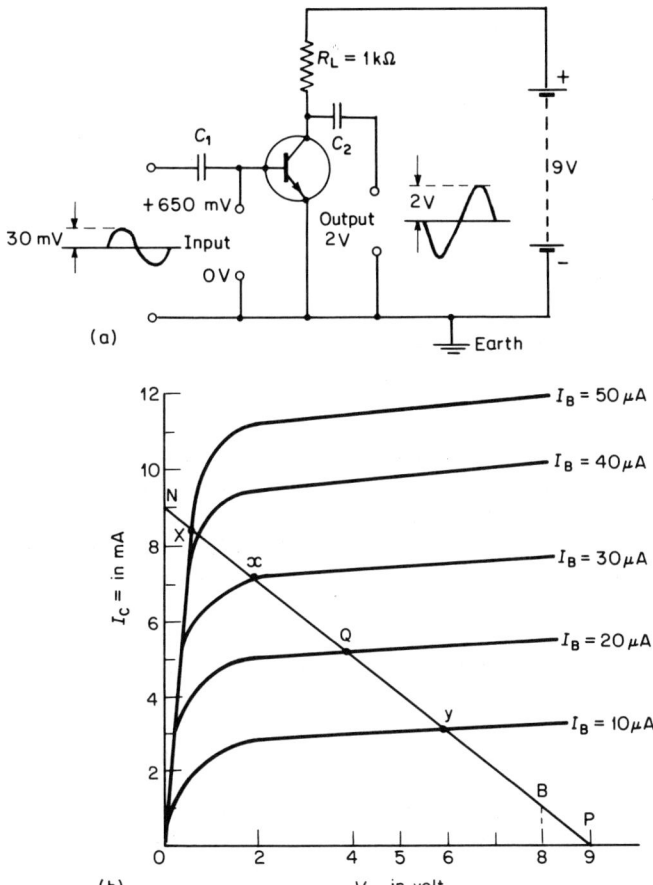

Figure 4.11. (a) An alternating voltage amplifier using an n-p-n transistor in common-emitter connection and (b) the load-line drawn on the I_C–V_{CE} characteristics

capacitor C_1 has an almost infinite resistance (because its dielectric is an insulator) to a d.c. supply.

It is an easy matter to select the value of C_1 as a passive component so that X_{C1} is negligibly small at the frequency f. This frequency is taken to be the lowest frequency of the alternating input signal likely to be encountered because X_{C1} is greater the smaller is f. The capacitor C_2 is needed to prevent the battery supply of 9 V from causing a direct current to pass via R_L through the output device. Again, the reactance X_{C2} of C_2 is made to be small at the lowest frequency of the alternating input signal so that the amplified version of this signal (due to the transistor) appears at the output but not the d.c. from the power supply.

The alternating input signal amplitude is only 30 mV; it is a small signal input of value considerably less than the operating d.c. input bias used. This d.c. bias of $+650$ mV on the base with respect to the emitter decides the operating point Q on the I_C against V_{CE} characteristics for a mean base current, taken to be 20 μA in the present case (Figure 4.11(b)).

The applied alternating signal causes the base voltage to vary sinusoidally between $+680$ mV and $+620$ mV. The corresponding base current variations are taken to be between 30 μA and 10 μA.

A load-line NQB is drawn as a straight line through the operating point Q across the $I_C - V_{CE}$ characteristics. This load-line represents the linear relationship between current and voltage for the particular load resistance R_L used.

Suppose, for example, that R_L is 1 kΩ. To construct the load-line, consider that when V_{CE} is equal to zero, all the applied voltage must appear across R_L, i.e. the voltage drop across R_L is 9 V, the e.m.f. of the battery supply. For 9 V across 1 kΩ, the collector current must be 9 mA. Hence point N on the load-line (Figure 4.11) must have coordinates $V_{CE} = 0$ and $I_C = 9$ mA.

When I_C is equal to zero, all the applied voltage must appear across the transistor, that across R_L being zero. Hence $V_{CE} = 9$ V and $I_C = 0$ are the co-ordinates of point P. The load-line is thus the straight line which joins N and P.

The main purpose of this load-line is to indicate the limits between which the amplifier can be operated without distortion, i.e. so that the waveform of the output signal is a true replica of that of the input signal. This freedom from distortion will be preserved provided that the sections xQ and Qy on the load-line are equal in length, where the successive $I_C - V_{CE}$ characteristics drawn are for equal increments of the base current I_B.

With the operating point chosen at Q and with the input signal causing the base current to vary between 10 and 30 μA, the load-line indicates that V_{CE} varies between $+2.0$ V and 6.0 V. Hence an output signal voltage of 4.0 V peak-to-peak is obtained.

Distortion of the output signal will occur if the variation in the base current takes this signal beyond the linear region of operation defined by the points x and y. As the voltage across the load approaches 9 V, if the base current increases further, the collector current cannot follow. This condition occurs as the operating point moves to the point X. The transistor is then said to be *saturated* or *bottomed*. Beyond the point B on the load-line, when the collector-current is very small, distortion will occur as the transistor approaches *cut-off*.

This graphical approach shows that the amplifier will have the following features:

(a) Current gain $\quad A_i = \Delta I_C/\Delta I_B = 4 \text{ mA}/20 \text{ } \mu\text{A} = 200.$

(b) Voltage gain $\quad A_v = \Delta V_{CE}/\Delta V_{BE} = 4.0 \text{ V}/60 \text{ mV} = 67.$

(c) Power gain $\quad A_i A_v = 200 \times 67 = 1.34 \times 10^4$

4.7 Circuit Analysis by Means of a Simple Equivalent Circuit

The alternating current behaviour of any bipolar transistor amplifier can be calculated by use of the hybrid parameters together with a circuit which is a simple model of the amplifier. Dealing with common emitter connection, the parameters which will determine the behaviour of the amplifier are the input resistance h_{ie}, the output resistance $1/h_{oe}$ and the forward current transfer ratio h_{fe}. If the output resistance is high (at least ten times the load resistance) the amplifier can be represented by the equivalent circuit of Figure 4.12.

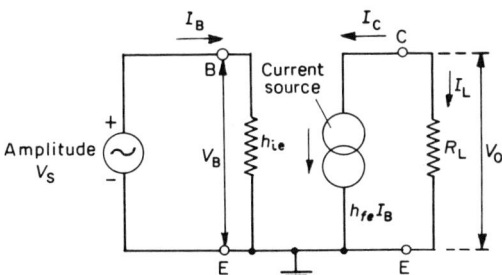

Figure 4.12. Simple equivalent circuit of a common-emitter amplifier

The equivalent circuit is derived on the basis that a sinusoidal input signal of amplitude V_s is applied between the base and the emitter and causes a peak base current of I_B to flow through the input resistance h_{ie}. The output circuit represents the behaviour of the transistor as a source of constant current $h_{fe}I_B$ (i.e. the forward current transfer ratio times the peak base current) which flows through the load resistance R_L to produce an output voltage of peak value V_O.

Example 4.7

A silicon n-p-n bipolar transistor (e.g. a BC107) is connected into the common-emitter amplifier stage shown in Figure 4.13. The steady emitter current I_E = 0.45 mA. If the transistor parameters are $h_{fe} = 200$ and $h_{ie} = 11.5 \text{ } k\Omega$.

(a) *Determine the steady voltages V_E, V_C and V_B and the steady currents I_B and I_{BB}.*

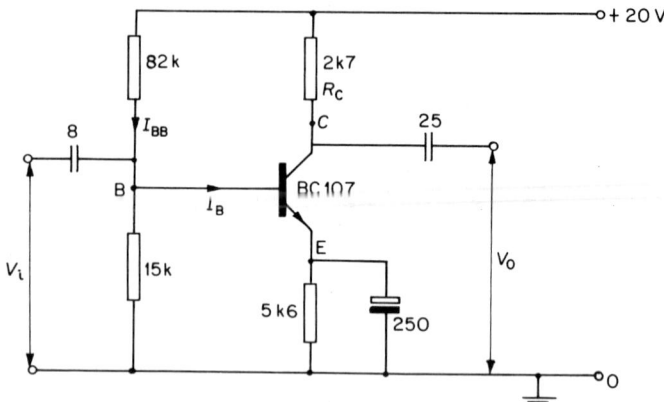

Figure 4.13. Concerning Example 4.7: a common-emitter amplifier stage based on an n-p-n transistor of type BC107*

(b) *Draw an approximate a.c. equivalent circuit for this amplifier stage, assuming that* $1/h_{oe} > 10R_C$,

(c) *Use this equivalent circuit (part b) to calculate the input resistance of the stage.*

(d) *Estimate the voltage gain of the stage.*

(e) *If the 250 μF capacitor were disconnected, estimate the voltage gain of the stage.*

(a)
$$V_E = 5.6 \times 10^3 \times 0.45 \times 10^{-3} = 2.5 \text{ V}$$

$$V_C = 20 - (2.7 \times 0.45) = 18.8 \text{ V}$$

$$V_B = 2.5 + 0.6 = 3.1 \text{ V}$$

Alternatively, $20 \times 15/97 = 3.1$ V

$$I_B = I_E/h_{fe} = 0.45 \times 10^{-3}/200 = 2.25 \ \mu\text{A}$$

Assuming that this value of I_B is much less than I_{BB} (if it turns out that it is not, a new approach to the calculation is needed because Kirchhoff's second law must certainly apply at point **B** in the circuit diagram).

$$I_{BB} = 20/(97 \times 10^3) = 206 \ \mu\text{A}$$

which is comfortably many times I_B.

* Note here the use of an elongated rectangle instead of a zig-zag line as the circuit diagram symbol for a resistor. Also, 82 k, for example, means 82 kΩ. Note that 8 placed against the symbol for a capacitor means 8 μF and, likewise, 250 means 250 μF. In the latter case, the symbol drawn is for an electrolytic capacitor for which the top plate (drawn open) is conventionally positive in potential with respect to the other plate (drawn full black).

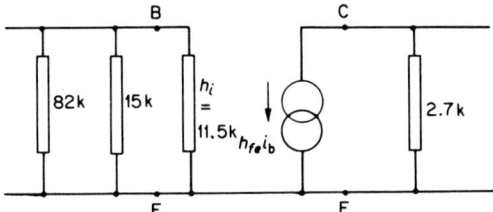

Figure 4.14. The a.c. equivalent circuit concerned in Example 4.7(b)

(b) The a.c. equivalent circuit is given here as Figure 4.14.

(c) The input resistance R_i is given by:

$$1/R_i = 1/82 + 1/15 + 1/11.5$$
$$= 0.0122 + 0.0666 + 0.0869 = 0.1657$$

so that

$$R_i = 1/0.1657 = 6 \text{ k}\Omega$$

(d) $A_v = h_{fe}i_b R_C / i_b h_{ie} = h_{fe} R_C / h_{ie}$

$$-200 \times 2.7/11.5 = -47$$

(e) Removing the 250 μF capacitor, then $i_c = i_e$, approximately and the gain $= R_C/R_E = -2.7/5.6 = -0.48$.

4.8 Biasing the Bipolar Junction Transistor

In the design of a bipolar amplifier based on the common-emitter circuit the steady value of the collector-to-emitter voltage and of the emitter current (i.e. V_{CE} and I_E, respectively) must be fixed and it must be arranged for the transistor to operate about the working point Q (Figure 4.11) which is maintained despite temperature fluctuations. The biasing circuit must therefore not only stabilize the working point it must also prevent thermal runaway*. This is much more likely to occur in germanium than in silicon transistors. A number of biasing methods for a bipolar transistor in common-emitter connection are described.

(a) *The fixed bias circuit* A resistance R_B is connected between the supply voltage terminal and the base (Figure 4.15). The base current I_B required to establish the best working point is known from the

---

* The leakage current between the collector C and the base B is I_{CBO}, with the emitter open, circuit. So, $I_{CEO} \simeq h_{FE}I_{CBO}$, where I_{CEO} is the total leakage current in the collector circuit. A rise in temperature will increase the leakage current and shift the operating point, so causing the amplified signal to be distorted. Also, the leakage current causes a temperature rise which results in increased leakage current. The process is thus cumulative and is called *thermal runaway*.

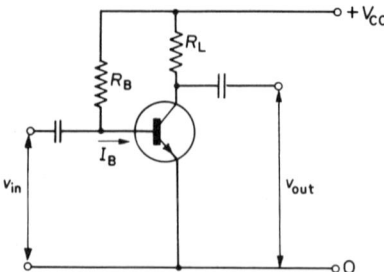

Figure 4.15. The fixed bias circuit

characteristics (Figure 4.11(b)). It is then a simple matter to calculate R_B because

$$I_B = V_{CC}/R_B \qquad (4.1)$$

assuming V_{EB} to be negligible, V_{CC} being the battery supply voltage. This circuit is thermally unstable: any increase in temperature which causes the leakage current I_{CEO} to increase will increase the collector current. Even if thermal runaway does not occur, distortion is likely at elevated temperatures.

However, a thermal runaway rarely occurs with modern silicon bipolar transistors so the effect is not so important as it was when germanium was the dominant semiconductor. The effect of elevated temperature (unless excessive) is also less important with silicon transistors.

(*b*) *Collector-to-base bias*   An improvement in stability is obtained if the resistance R_B is connected between collector and base (Figure 4.16). If the collector current tends to increase, the collector voltage decreases, hence I_B decreases so tending to reduce the collector current; this is d.c. negative feedback. Neglecting the voltage V_{BE}, the

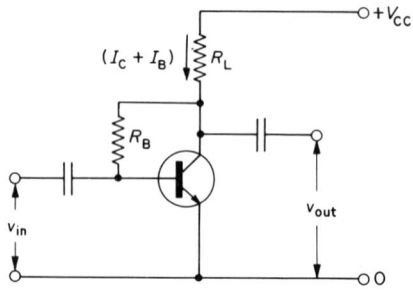

Figure 4.16. Collector-to-base bias

required value of the base resistance R_B is calculated, knowing V_{CC}, I_C, I_B and R_L, from

$$I_B = \frac{V_{CC} - I_C R_L}{R_L + R_B} \qquad (4.2)$$

One fundamental disadvantage of this circuit is that part of the alternating output signal which is 180° out of phase with the input signal is fed back into the input circuit via R_B. This phenomenon is a.c. negative feedback (see section 4.10). The d.c. negative feedback aids the stability; the a.c. feedback reduces the gain of the stage.

(c) *Collector-to-base bias with the base resistor decoupled* The reduction of gain due to the a.c. feedback resulting in case (b) may be eliminated by dividing R_B into two equal parts and connecting the junction between these resistors to the common-emitter *via* a capacitance C_1 which has a negligible reactance at the signal frequencies (Figure 4.17). The base resistance R_B is said to be decoupled* at the signal frequencies. The alternating output signal is now not fed back into the input circuit.

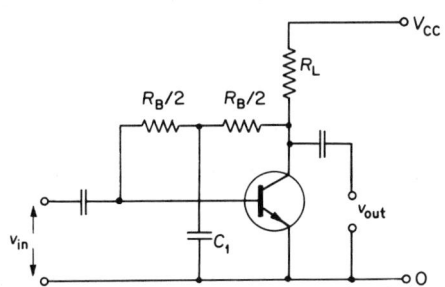

Figure 4.17. Collector-to-base bias with R_B decoupled

Collector-to-base biasing is often used, especially with silicon transistors in which the leakage currents are low.

(d) *Potential divider and emitter resistor stabilizing bias* The necessary steady bias on the base is provided by a potential divider consisting of resistances R_1 and R_2 across the battery supply voltage V_{CC} (Figure 4.18). Furthermore, the emitter is not connected directly to the (often earthed) terminal of the supply — the negative terminal

---

* The term 'decoupled' is widely used in electronics: it is always done with a capacitor C. This capacitor acts as a low reactance to the flow of alternating current at the signal frequency but as a virtually infinite resistance to direct current. In Figure 4.17, the capacitor C_1 effectively short-circuits to earth the junction point between $R_B/2$ and $R_B/2$.

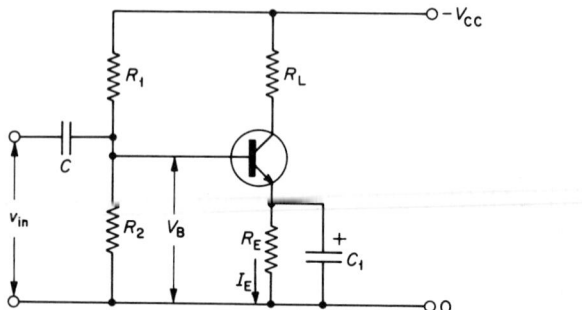

Figure 4.18. Bias with a potential divider and emitter resistor stabilizing

for an n-p-n transistor — but via a series resistance R_E. This resistance R_E carries the emitter current I_E and so there is a steady bias on the emitter relative to earth of $I_E R_E$. When an alternating signal is applied to the input via the capacitance C, there will also be an alternating component of the emitter current. To ensure that this does not introduce a corresponding alternating bias on the emitter, the capacitance C_1 is connected in parallel with R_E. The reactance of C_1, which is $1/(2\pi f C_1)$ where f is the input signal frequency, is chosen by virtue of the magnitude of C_1, to be much smaller in magnitude than R_E.

With this provision of R_E, any increase in collector current causes an increased voltage drop across R_E so that V_{EB} decreases, causing a smaller base current to flow. This in turn means that I_E will increase less than it would have done had there been no self-biasing resistance R_E.

By choice of R_1 and R_2, the current through them is arranged to be several times the base current I_B. Consequently,

$$V_{BE} = V_B - I_E R_E \tag{4.3}$$

where V_B is the voltage drop across R_2. Hence

$$I_E = V_B/R_E \quad \text{approx.} \tag{4.4}$$

because V_{BE} is small.

The steady emitter current or the collector current ($I_E \simeq I_C$) can be calculated from equation (4.4) merely from a knowledge of the resistance values forming the potential divider and the value of R_E. This method of obtaining bias by potential divider and emitter resistor stabilizing is often used.

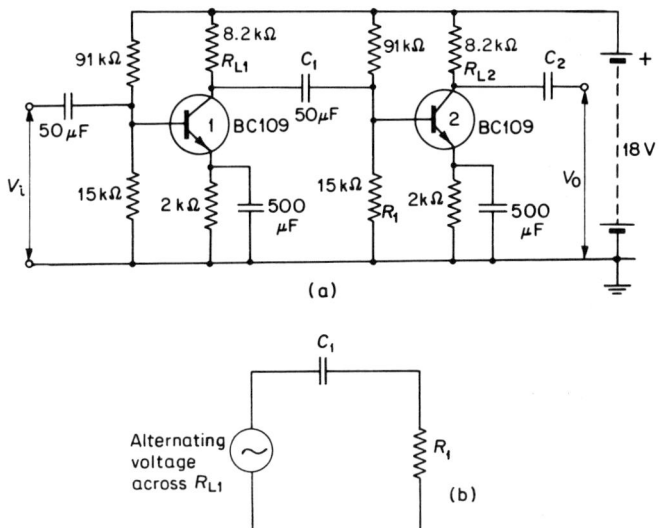

Figure 4.19. A two-stage, alternating voltage amplifier

4.9 A Two-Stage Common-Emitter Transistor Amplifier

The most frequently used method of coupling together two stages in an alternating voltage amplifier is by resistance-capacitance coupling (Figure 4.19(a)). This circuit uses two n-p-n junction transistors (BC109) with potential divider, emitter resistor stabilizing bias, and the component values given are suitable for alternating voltages of audio-frequency over the range from 20 Hz to 30 kHz.

To indicate how the alternating voltage which appears across the load resistance R_{L1} of stage 1 (which is an amplified replica of the input to the amplifier) is connected to the input of the second stage 2, consider that a capacitance C_1 and resistance R_1 are in series across R_{L1} (ignoring* other components for simplicity). It is seen from Figure 4.19(b) that the alternating voltage across R_1 is $R_1/[\sqrt{R_1^2 + (1/2\pi f C_1)^2}]$ for a frequency f. If C_1 is large enough for $(1/2\pi f C_1)$ to be much small than R_1, almost the whole of the alternating voltage across R_{L1} appears across R_1, forming the input to stage 2.

In an experiment, the alternating input voltage V_i can be conveniently obtained from a transistor oscillator with a potential divider across its output terminals to ensure that an input signal is obtained which is small enough not to overload the amplifier and cause distortion. This input voltage can be measured by either a multi-range a.c. digital voltmeter or a calibrated cathode-ray oscillograph. The output voltage V_0 from the amplifier is across R_{L2},

---

* This includes the 91 kΩ resistor in parallel (in effect) with R_1.

conveniently isolated from the steady voltage by the decoupling capacitance C_2 so that V_0 appears across the collector (except for the low reactance of C_2) of stage 2 and earth. V_0 is also measured with the digital voltmeter or the cathode-ray oscillograph.

If the oscillator used to provide the input is of suitably variable frequency, the overall voltage amplification A_v provided by the two-stage amplifier may be plotted against the frequency over the range, say, from 20 Hz to 30 kHz.

4.10 Feedback

Feedback in an amplifier occurs when the output exerts some influence on the input. This may occur fortuitously or it may be introduced deliberately. When the signal fed back from the output to the input is 180° out of phase (in anti-phase) with the input signal, the amplification is decreased and the feedback is said to be *negative* or *degenerative*. On the other hand, when the signal fed back is in phase with the input signal, the amplification is increased and the feedback is *positive* or *regenerative*.

It would seem at first that positive feedback is desirable. However, it is practically never used in amplifier design because its effect is cumulative leading to instability. Positive feedback increases the input signal which in turn increases the output signal which then increases the input still further. Though not used in amplifiers it is nevertheless employed in oscillators.

Negative feedback is always used in the design of high quality amplifiers. Though it must clearly result in loss of amplification because part of the output is fed back to reduce the input, a number of distinct advantages accrue from its use which far outweigh this reduced amplification. In any case, further amplification can be obtained by simply adding another stage.

The conventional block symbol for an amplifier is a triangle (Figure 4.20) in which the letter A is inscribed to denote its gain. The signal from input to output is imagined to travel in the direction of the

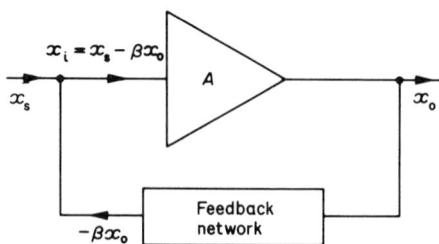

Figure 4.20. Block diagram to illustrate negative feedback

apex of the triangle: the input is therefore at the base; the output is at the apex.

Let the actual initial signal be x_s and the output signal that this would give be x_o. Suppose that by some means a fraction β of this output is fed back to the input circuit so as to be 180° out of phase with the input, i.e. negative feedback is practised. The input signal now becomes x_i given by

$$x_i = x_s - \beta x_o$$

Before feedback was applied, the input was x_s and the output was $x_o = x_s A$. With negative feedback applied, the signal $-\beta x_o$, equal to $-\beta x_s A$, is introduced into the input. Hence to obtain the same output as before the input signal must be increased by $\beta x_s A$, to become $x_s(1 + \beta A)$. The gain of the amplifier with negative feedback applied is therefore A_f given by

$$A_f = \frac{x_o}{x_s(1 + \beta A)} = \frac{A}{1 + \beta A} \tag{4.5}$$

Although the fraction β (known as the *feedback factor*) may be small, A can be very large so that the term βA is large compared with unity. In this case, 1 is negligible in the denominator of equation (4.5) so that

$$A_f = \frac{A}{\beta A} = \frac{1}{\beta} \tag{4.6}$$

This equation is of paramount importance in negative feedback. It shows that the gain of the amplifier with negative feedback depends almost entirely on the circuit arrangement which decides β and is largely independent of A (often called the *open-loop gain* of the amplifier, i.e. the gain with zero feedback).

In a simple case, therefore, the gain is virtually independent of the transistor used. If a transistor becomes faulty and is replaced by another one of the same type but different characteristics (and the characteristics of transistors of the same type may vary considerably because of difficulties in maintaining a consistent semiconductor manufacturing process) the amplifier circuit behaviour is unchanged.

Other benefits resulting from the use of negative feedback are:

(a) very stable operation;
(b) low distortion;
(c) reduction of noise;
(d) the ability to control the input and output resistance of the circuit.

Further discussion of these factors is beyond the scope of this text.

The remainder of this chapter is concerned with various useful simple but basic circuits amongst the very large number that can be constructed using junction transistors.

Example 4.10(a)

An amplifier with an open-loop gain of 50 is used in a negative feedback amplifier system. If 15 per cent of the output signal is fedback in anti-phase to the input signal, determine the input signal voltage needed to obtain an output of 1 V.

As the closed-loop gain $A_f = A/(1 + \beta A)$ where A is the open-loop gain, therefore

$$A_f = 50/(1 + 15 \times 50/100) = 50/8.5$$

So

$$v_o/v_i = 50/8.5 = 1/v_i$$

Therefore

$$v_i = (8.5/50)\text{V} = 170 \text{ mV}$$

Example 4.10(b)

An amplifier with an open-loop gain of 2000 is used in a negative feedback amplifier system. If 10 per cent of the output signal is fedback in anti-phase determine the gain of the system. The open-loop gain of the amplifier drops by 10 per cent during its working life. Calculate the new gain of the system. Comment.

As the closed-loop gain $A_f = A/(1 + \beta A) = 2000/(1 + 0.1 \times 2000)$

$$A_f = 2000/201 = 9.95$$

The open-loop gain drops to 1800 now

$$A_f = 1800/(1 + 0.1 \times 1800)$$

$$= 1800/181 = 9.94$$

Comment: there is a very small difference between these last two values of A_f, as would be expected.

4.11 A Simple Constant-Current Source

A source which provides a constant current to a load is frequently required in laboratory practice. A simple example is the necessity for the current through a filament lamp to be constant if the light output from the lamp is to be unvarying.

Ideally a constant current source should have an output resistance ($\Delta V/\Delta I$, where ΔV is the change of voltage for a current change of ΔI) which is very large (ideally, infinite) so that the introduction of any load resistance R_L has very little effect on the current that flows through it. Thus if R_L varies, its effect is insignificant on $\Delta V/\Delta I$ (because R_L is much less than $\Delta V/\Delta I$) and so on the current I which flows through it.

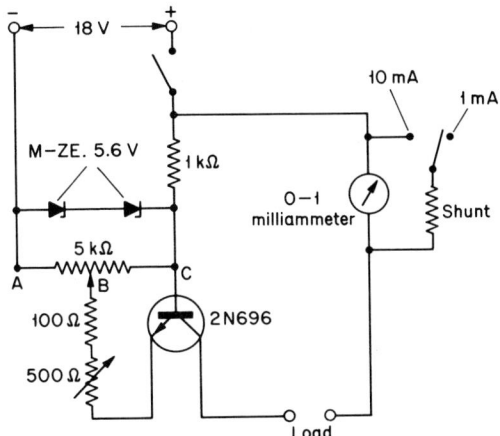

Figure 4.21. A constant-current source

A transistor in common-base connection has a very high output resistance, $1/h_{ob} = \Delta V_{CB}/\Delta I_C$ (section 4.3). However, its emitter-base voltage must be kept constant. In the constant-current circuit (Figure 4.21), therefore, the silicon n-p-n transistor used has its emitter-base voltage supply from a Zener diode stabilized supply so that the p.d. across the 5 kΩ resistor is 11.2 V at all times because of the two Zener diodes (Type M-ZE which each stabilize at 5.6 V) in series across it. The 5 kΩ potentiometer provides a coarse control and the variable 500 Ω resistor a fine control for the current required.

Some negative d.c. feedback is present. This is because, if the collector current I_C of the n-p-n junction transistor (type 2N696) increases for some reason, the emitter current I_E also increases. Hence the p.d. across the section AB of the 5 kΩ resistor increases and so that across the section BC (which decides the emitter-base voltage on the transistor) must fall; this tends to reduce I_E. The resistances in series in the emitter circuit of the transistor also reduce the emitter-base voltage if I_E increases.

The collector-base voltage supplied to the transistor is that across the 1 kΩ resistor. It is 6.8 V in the circuit of Figure 4.21. Negligible current change (< 0.5 per cent) can be detected by the milliammeter in the output circuit of the transistor as the load resistance is altered, until the transistor 'bottoms' or 'saturates'. This will occur only when all the 6.8 V is dropped across the load resistance.

An experiment based on such a circuit can be undertaken to show that for a current of 10 mA through the load, the change of this current is less than 0.5 per cent for variations of the load between 0 and 600 Ω. If the current maintained through the load is 1 mA, this load can be varied from 0 to 6000 Ω with less than 0.5 per cent load current change, and with the current at 0.5 mA, the load can be altered between 0 and 12 000 Ω. This constant current source is of particular value in the operation of the following apparatus:

(i) To measured the variation of the conductivity of germanium with temperature (section 1.17).

(ii) To examine the characteristics of a thermistor (section 5.2). The small current passed through the thermistor produces negligible heating but enables the resistance to be measured at a number of temperatures simply by using a high resistance voltmeter.

(iii) To operate Hall probes.

4.12 Stabilized Power Supplies

The basic Zener diode regulator circuit (section 3.15) enables a supply of voltage to be stabilized (section 3.16) and is frequently used when the current demand is small. The limitation is the power-handling capacity of the Zener diode. In general, if the current demand exceeds 50 mA, or if increased stability is needed, or if a variable output voltage is required, a series transistor voltage regulator is used.

The stable Zener voltage is applied between the base and the emitter of the n-p-n bipolar transistor (Figure 4.22). The p.d. across

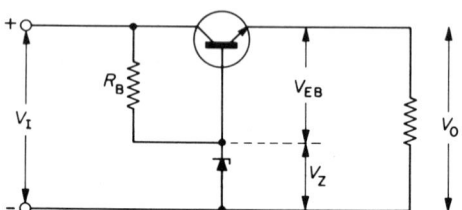

Figure 4.22 A series transistor regulator

the resistor R_B provides the collector-base voltage. The output voltage V_o is given by

$$V_o = V_Z - V_{EB} \tag{4.7}$$

As V_{EB} is small, approximately 0.6 V for a silicon transistor, $V_o \simeq V_Z$, where V_Z is the constant p.d. across the Zener diode.

If the input voltage V_I (usually that of a power pack) changes for any reason (e.g. a.c. mains supply fluctuations) the collector-base voltage V_{CB} (across R_B) changes by the same amount because $V_i = V_{CB} + V_Z$ and V_Z is constant. The series control transistor is said to be operating in 'emitter follower' connection, because any voltage applied between its base and emitter is reproduced across any load in its emitter lead. The output voltage V_0 remains unchanged however, because the change of V_{EB} is negligibly small.

For good regulation, the d.c. current gain h_{FE} of the transistor used should be large and the current through the Zener diode should be large compared with the base current of the transistor.

A useful circuit for operating semiconductor electronic circuits provides a constant 10 V able to supply currents of up to 40 mA (Figure 4.23). The subminiature mains transformers used with a

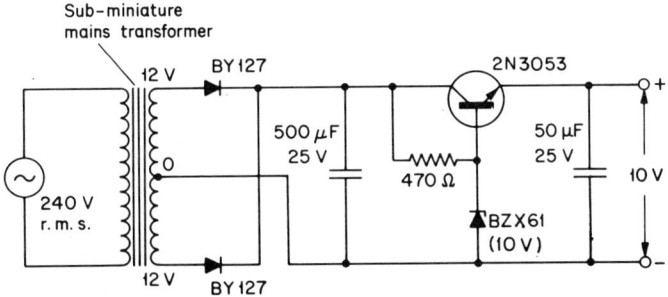

Figure 4.23. A 10 V stabilized voltage supply

secondary providing $12 - 0 - 12$ V is a most convenient component for use in a small power supply because it fits neatly on to a circuit board with the other components. However, its mains input terminals are not shrouded so that some thought should be given in construction to avoid the hazard of exposed 240 V terminals. Though simple to construct and providing a useful laboratory unit, this power supply has two main disadvantages:

(a) No provision exists for varying the output voltage which is almost equal to V_Z, the voltage of the reference Zener diode BZX61 (10 V).

(b) Changes in V_{EB} and V_Z due to changes in temperature appear at the output.

Note that the series transistor which, in effect, absorbs any voltage variations, must be capable of dissipating the power produced at its junctions. It may therefore required to be mounted on a heat sink. For the 2N3053 transistor of Figure 4.23, which is within a metal can termed a TO39 (transistor outline) configuration, a SISTASINK type 2215 heat sink is ideal.

The power produced within the transistor is not necessarily a maximum when the current through it is a maximum. The power is most readily calculated approximately from the product of the load current I_L and the voltage V_{CB} across the reverse-biased collector-base junction (i.e. $I_L V_{CB}$).

The stabilization of voltage supplies is considered again in Chapter 7 because the use of differential operational amplifiers has introduced significant changes to the treatment of this important subject.

4.13 Logarithmic Units for Power Ratios: the Decibel

It is convenient to express power, voltage and current ratios on a logarithmic scale, known as the decibel scale. The ratio of two powers P_1 and P_2 expressed in bels is given by $\log_{10}(P_2/P_1)$. In decibels (dB) this is

$$10 \log_{10}(P_2/P_1) \tag{4.8}$$

Very often P_1 is chosen to be some reference power level.

For an amplifier, P_2 would be the output power and P_1 the input power. If the input and output resistances of an amplifier are both equal to R, $P_2 = V_2^2/R$ and $P_1 = V_1^2/R$, where V_2 is the output voltage and V_1 is the input voltage. Under these circumstances, (4.8) can be written

$$10 \log_{10}[(V_2^2/R)/(V_1^2/R)] = 20 \log_{10}(V_2/V_1)$$
$$= 20 \log_{10} A_V \tag{4.9}$$

Expressed in terms of the input current I_1 and the output current I_2, (4.8) becomes

$$10 \log_{10}[(I_2^2 R)/(I_1^2 R)] = 20 \log_{10}(I_2/I_1)$$
$$= 20 \log_{10} A_I \tag{4.10}$$

It is common practice — even though it is somewhat misleading — to ignore the fact that the output and input resistances are generally not equal, and simply to express the voltage gain of an amplifier in dB by $20 \log_{10}(V_o/V_i)$ where V_o is the output voltage and V_i is the input voltage.

When a number of amplifier stages in a multi-stage amplifier, each of known gain are coupled together the overall gain is the product of the individual gains. If these gains are expressed on a logarithmic scale, the overall gain is simply the sum of the individual gains.

Voltage ratio	Power ratio	dB
1	1	0
1.414	1.995*	3
1.995*	3.98†	6
3.162	10	10
10	100	20
31.62	1000	30
100	10^4	40
1000	10^6	60

* Approximately 2; † approximately 4.

Table 4.1. Voltage and power ratios expressed on the dB scale.

A number of voltage and power ratios are expressed on the dB scale in Table 4.1.

A negative value for the gain expressed in dB means that the output power is below the reference level or that the output voltage of an amplifier is less than the input voltage. For example, if the output voltage of the amplifier is 0.1 of the input voltage, $20 \log_{10}(0.1) = 20 \times \bar{1} = -20$.

An amplifier is usually designed to have a uniform voltage gain $A_v (= V_{out}/V_{in})$ over a wide frequency range. Such an amplifier is said to have a *flat response*. The band width of an amplifier is defined as $(f_2 - f_1)$ where f_2 is the upper frequency at which the power gain has dropped to one-half of its mid-band value, i.e. the power gain is '3 dB down' and the voltage gain has dropped to $A_v/\sqrt{2}$. Similarly f_1 is the lower frequency at which the voltage gain has dropped to $A_v/\sqrt{2}$.

4.14 Transistor Oscillators

An oscillator circuit is basically that of an amplifier in which a fraction of the output power is fed back to the input circuit so as to be in phase with the initiating input signal. This positive or regenerative feedback provides a self-generating current or voltage variation at a frequency dependent upon the component values used.

An RC Sinusoidal Oscillator

The output from an n-p-n transistor (e.g. type BC107) in common-emitter connection forming stage 1 is resistance-capacitance (RC) coupled (*cf.* section 4.9) to the input of a second similar transistor forming stage 2 (Figure 4.24). As each stage introduces a phase change of 180° (section 4.5) the resultant phase shift introduced by the two coupled transistors is 360°.

The load in the output of the second stage consists of two impedances in series: Z_1 comprising R_9 and C_3 in series and Z_2 comprising C_2 and R_{10} in parallel. The alternating voltage across Z_2 is made the input to the first-stage transistor. This fed-back input will be in phase with the output and so also in phase with the existing input provided that the alternating voltage across Z_2 is in phase with that across Z_1 and Z_2 in series. The feedback will then be positive so the circuit will oscillate. It can be shown that this zero phase shift occurs at a frequency f decided by

$$2\pi f = 1/\sqrt{(C_2 C_3 R_9 R_{10})} \qquad (4.11)$$

which is the frequency of oscillation.

It is usual to make $C_2 = C_3$ and $R_9 = R_{10}$ so that equation (4.11) becomes

$$f = 1/2\pi C_3 R_9$$

With $C_3 = 0.01$ μF and $R_9 = 4.7$ kΩ, as in Figure 4.24.

$$f = 1/(2\pi \times 10^{-8} \times 4.7 \times 10^3) = 3300 \text{ Hz approx.}$$

The output across resistance R_8 is of particularly good sinusoidal waveform.

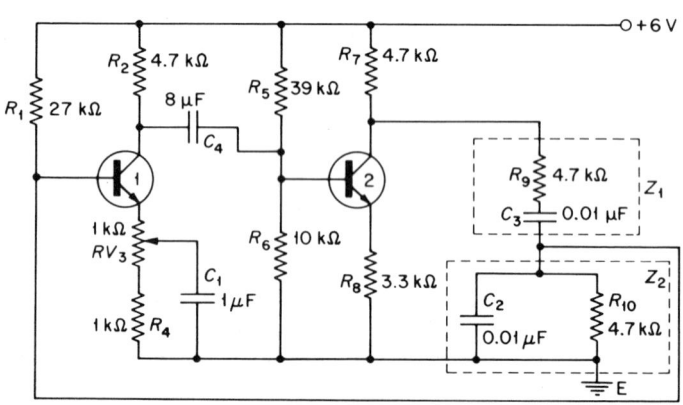

Figure 4.24. An *RC* sinusoidal oscillator

A Crystal-controlled Oscillator

If a slice of a piezoelectric crystal — usually quartz — has electrodes plated on opposite faces and a potential difference is applied between these electrodes, the crystal slice will become physically strained, i.e. a small change in its physical dimensions will occur. Conversely, if the slice is strained it becomes electrically polarized and free electric charges appear on the plated. faces.

If the slice is correctly cut with respect to the crystal axes, correctly mounted and made to vibrate at its natural frequency by the application of a specific resonant alternating p.d., a very stable frequency source is obtained. Slices of quartz are cut to dimensions for which the natural frequencies are within the range from a few kHz to a few MHz. A quartz crystal slice with a frequency of 465 kHz is common and readily available. Such a crystal is used in the oscillator shown in Figure 4.25.

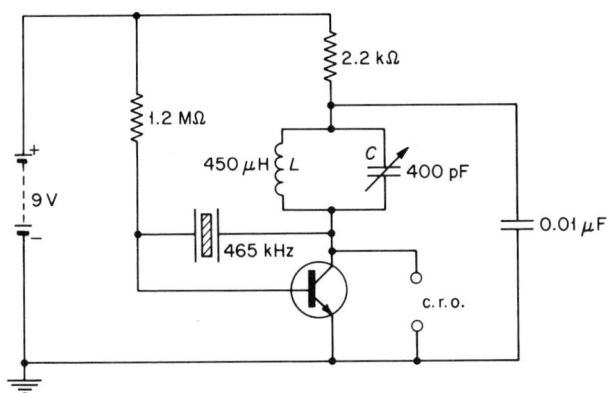

Figure 4.25. A crystal-controlled oscillator

The n-p-n transistor is operated with its emitter earthed and an inductance-capacitance (*LC*) circuit with a resistor of value 2.2 kΩ as a load in series between the +9 V terminal of the supply and the collector. The collector is coupled to the base of the transistor by the capacitance of the quartz crystal between its plates. The coupling provides positive feedback between the collector and the base. The 1.2 MΩ resistor is provided to ensure the correct d.c. base bias; the 2.2 kΩ resistor ensures the appropriate collector d.c. potential.

The tuned circuit *LC* does not determine the frequency of oscillation; it serves to vary (on altering the capacitance *C*) the amplitude of the oscillations which will be a maximum when the resonant frequency $1/[2\pi\sqrt{(LC)}]$ of this circuit and that of the crystal are the same.

The waveform of the oscillating voltage output is observed by means of a cathode-ray oscillograph joined between the collector and earth. This will be the same as that across the *LC* circuit because one end of this circuit is earthed via the 0.01 μF capacitor as regards a.c. and the other is joined to the collector.

A Phase-shift Oscillator

Consider a capacitance C and a resistance R in series. The alternating current through this combination leads in phase with the alternating voltage across it by an angle α given by

$$\tan \alpha = 1/\omega CR$$

so that

$$\alpha = \tan^{-1} (1/\omega CR),$$

where $\omega = 2\pi f$, f being the frequency of the alternating current.

At a particular frequency f, it is clearly possible to choose the values of C and R to be such that $\alpha = 60°$.* The use of three such stages in cascade can hence be used to provide an overall phase shift of 180°. If such an arrangement is used in conjunction with a transistor in common-emitter connection, it is possible for the 180° phase change introduced by the transistor itself (section 4.5) to be added to the 180° obtained by the use of the three RC stages to arrange that the output signal is fed back to produce an in-phase input signal so that an oscillator is obtained.

The circuit of a phase-shift oscillator (Figure 4.26) makes use of a high gain n-p-n transistor (for example, type BC107). The three phase shifting RC circuits consist of C_1R_1, C_2R_2 and C_3 in conjunction with the input resistance of the transistor, where $C_1 = C_2 = C_3 = 0.01$ μF

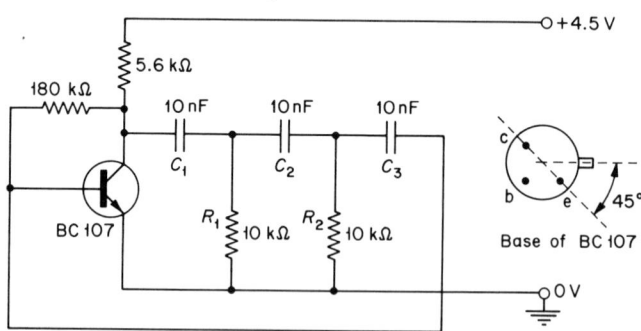

Figure 4.26. A phase-shift oscillator which makes use of an n-p-n transistor

---

*   A convenient value for explanation; in fact, the three RC sections do not have to be identical and rarely are. Indeed, for other RC networks, the frequency will adjust itself so that the part of the signal providing a phase shift of 180° will be further amplified in preference to any other.

and $R_1 = R_2 = 10$ kΩ. With a 4.5 V supply, a resistor of 180 kΩ arranges an appropriate positive bias on the base of the transistor which has its emitter connected to the earthed line. A load resistance of 5.6 kΩ is in series with the collector and the positive terminal of the supply. The voltage at the collector will vary in antiphase with any voltage variation on the base relative to the emitter. The three *CR* circuits introduce a further 180° phase change so that any voltage variation on the base is supplemented by an in-phase positive feedback.

A cathode-ray oscillograph connected between the collector and the emitter enables the waveform (which is sinusoidal) of the oscillating voltage output to be examined and its frequency to be determined.

4.15 Field Effect Transistors

The transistor already described is called a *bipolar* transistor because its performance depends on the interactions of two types of charge carrier: electrons and positive holes. The field effect transistor (FET) is a *unipolar* device because only one type of charge carrier is involved: electrons in an n-channel FET and holes in a p-channel FET. Only the n-channel FET (which is the more commonly used of the two) is described here. There is also a class of FET which is now probably more widely used than the original FET, especially for the logic circuits used in computers and microprocessors, it is the MOSFET (metal oxide semiconductor FET). Of this again both n-channel and p-channel types exist and, again, the concern is primarily with the n-channel type. The p-channel types of both the FET and the MOSFET behave in a way which is essentially similar to that of the n-channel versions. However, it must be remembered that the potentials supplied to the n-channel types are of the opposite polarities from those supplied to the p-channel models.

The problem which has been solved in the FET is to make a semiconductor device which has a very high input resistance, like the now obsolete thermionic vacuum tube (valve) but unlike the bipolar transistor which has a comparatively low input resistance. Thus FETs are essentially voltage input controlled devices whereas bipolars are essentially current controlled.

To arrive at this solution, the motion and concentration of the charge carriers within the semiconductor material of a FET are controlled by an electric field set up within the conducting region by voltages applied to contacts to the semiconductor material. Indeed, the idea of the FET preceded that of the bipolar transistor but early attempts to control the electric current within a bar of semiconductor by maintaining a voltage at or near the surface of the semiconductor were unsuccessful. In all cases, the surfaces appeared to screen electrically the interior of the semiconductor.

It was while investigating these surface effects that Shockley and his associates, Bardeen and Brattain, discovered transistor action in 1948 and eventually introduced the bipolar transistor.

Some years later Shockley was responsible for introducing a solid-state device with an input resistance of several hundred megohm. He realized that a controlling electric field could be created within the body of a semiconductor by the use of a reverse biased p-n junction. The device — often called a junction gate field effect transistor — depends for its action on the creation and control of the depletion layer produced at a reverse biased junction.

4.16 The Structure and Behaviour of an FET

The creation of a depletion layer has already been mentioned (section 3.3). Figure 4.27 shows the form of a depletion region for a p-n junction in which the semiconductor material in the p-type region has

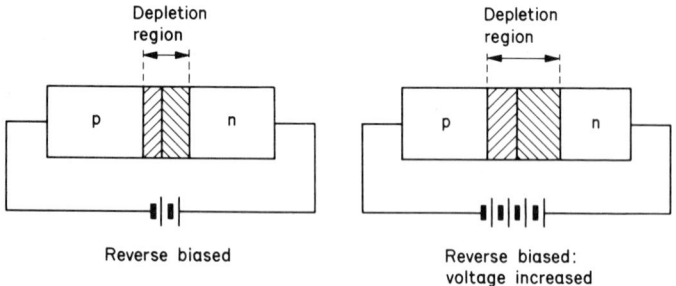

Figure 4.27. Effect of reverse bias on a depletion region about a p-n junction where the p-type region is more heavily doped

a much greater carrier concentration (is more heavily doped) than the material in the n-type region.

As the reverse bias is increased, the width of the depletion region increases, but it always extends further into the n-type material than into the p-type. This is because each type of material must contribute the same number of current carriers so that a greater volume of the less heavily doped n-type material is depleted. On the n-type side of this junction, the loss of electrons which have diffused across the junction leaves positive ions; on the p-type side, loss of holes which have diffused across the junction leaves negative ions. The ions are locked in the crystal lattice: they are not mobile. The conductivity of this depletion region is nominally zero because mobile current carriers are not available.

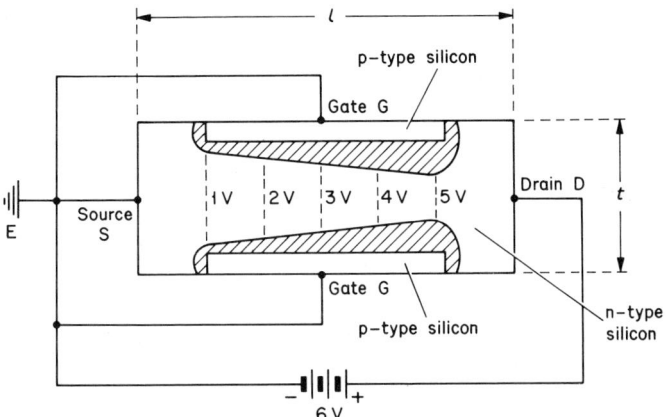

Figure 4.28. Simple model of the action of an n-channel FET

A simple model of an n-channel FET may be envisaged by considering a bar of n-type silicon with ohmic contacts at each end (Figure 4.28). If this bar has length l, width w and thickness t, the resistance R between these end contacts is given by

$$R = \rho l / wt$$

where ρ is the resistivity of the n-type silicon. A battery providing an e.m.f. of, say, 6 V is connected across the ends of this bar where the ohmic contact electrode at the negative end (sometimes earthed) is called the *source* S and the contact at the positive end is the *drain* D. The voltage across D and S, V_{DS}, is called the drain-to-source voltage.

Within each side of this bar of thickness t is produced a layer of heavily doped p-type silicon. The electrodes GG attached to these layers are called the *gate*. Normally this gate is biased negatively with respect to the source S, the gate-to-source voltage being V_{GS}, where V_{GS} is negative.

The current to the drain electrode, I_D, is due to the movement of electrons along the n-type silicon bar. The current to the gate electrodes, I_G, is vanishingly small if V_{GS} is negative because no mobile current carriers are available to cross the reverse-biased p-n junctions. The drain current I_D is a function of both the drain-source voltage V_{DS} and the gate-source voltage V_{GS}.

The important characteristics of the FET are hence the drain current I_D plotted against the drain-source voltage V_{DS} for various fixed values of the gate-source voltage V_{GS}. A given value of V_{GS} which is negative will produce a reverse bias across the two p-n junctions on either side of the silicon bar of which the widths of the depletion regions depend on the magnitude of V_{GS}. An increase from zero of the drain-source voltage V_{DS} (with V_{GS} constant) will have initially a two-fold effect:

(a) it will cause the drain current I_D to increase on the presumption that the n-type silicon bar is simply an ohmic conductor so the velocities of the electrons will increase;

(b) it will also increase the width of the depletion regions about the p-n junctions associated with the gate electrodes: this will cause I_D to decrease.

The statement (b) needs to be examined further. Suppose that V_{DS} is 6 V. Ignoring for the moment the gate-source voltage V_{GS}, it is clear that there will be a distribution of voltage along the length of the silicon bar due to V_{DS}. This distribution will increase from zero at the source to 1, 2, 3 etc. up to 6 V at the drain (Figure 4.28). The voltages all act as reverse bias values adding to the gate-source voltage, making it effectively more negative. The maximum negative bias, and so the maximum depletion region width, will exist over a small region near the drain-end of the gate electrodes.

This increase of the depletion region widths reduces the effective conducting cross-section of the n-type silicon bar. A *pinch-off effect* is obtained: the electrons cannot travel so readily to the drain through this narrow channel, so the drain current I_D decreases.

With a given value of V_{GS}, the effect of increasing the drain source voltage V_{DS} is a consequence of both actions (a) and (b). The result is that the drain current I_D increases at first almost linearly as V_{DS} is increased from zero. When V_{DS} is a few volts positive, however, the effect (b) becomes more and more pronounced, resulting in the fact that I_D becomes independent of a further increase of V_{DS}. Thus, beyond a certain value of V_{DS}, I_D is constant.

It is not easy to prove that I_D becomes a constant independent of V_{DS}. This can be shown to be the case by a more intimate study of the voltage gradient in the vicinity of the narrowest part of the channel between the depletion regions, but involves a study of the mobility of the electrons which becomes a function of the electric field strength (voltage gradient) at high values and will not be attempted here.

The effect on the drain current I_D of increasing the negative bias on the gate, V_{GS}, for a given value of the drain-source voltage V_{DS} is that it increases the width of the depletion regions around the p-n junctions. This causes I_D to decrease. If V_{GS} is made sufficiently negative, the pinch-off effect of these depletion region widths is great enough nominally to extend right across the silicon bar and reduce I_D to zero.

The terms which have been used to relate specifically to the FET are:

(i) *The source S*: the electrode by which the majority carriers enter the bar.

(ii) *The drain D*: the electrode by which the majority carriers leave the bar.

(iii) *The gate G*: the heavily doped p-region (often denoted by p^+, see section 2.4) is called the gate electrode because of its controlling function.

(iv) *The channel*: the region in the bar between the two gate electrodes. The majority carriers move through this channel from the source to the drain. Electrons are the majority carriers in an n-channel FET (one based upon an n-type silicon bar), whereas positive holes are the majority carriers in a p-channel FET.

Because the mobility of holes in silicon is only half that of electrons, the p-channel FET has inferior characteristics to an n-channel one of the same geometry. Consequently, n-channel FETs are the much more widely used.

In Figure 4.29 are shown the circuit symbols for an n-channel FET, a p-channel FET and a practical FET structure. Certain polarity

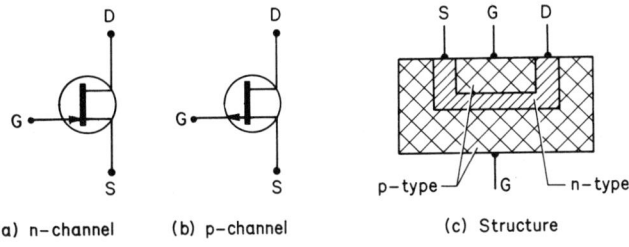

(a) n-channel   (b) p-channel   (c) Structure

Figure 4.29. Circuit symbols and structures of an FET

conventions are illustrated by Figure 4.30. If the polarity of the current or voltage is opposite to that indicated by the arrow, the value noted must carry a negative sign. Hence an n-channel FET operating in a normal manner might have a drain-source voltage V_{DS} of $+20$ V, and a drain current I_D of $+1$ mA with a gate-source bias V_{GS} of -2 V. On the other hand, a p-channel FET operated under similar conditions would have $V_{DS} = -20$ V, $I_D = -1$ mA and $V_{GS} = +2$ V.

4.17 An Experiment to Plot the Drain Characteristics of an n-channel FET

In the circuit employed (Figure 4.31(a)) the drain-source voltage V_{DS} can be set and recorded at values between 0 and 25 V, the drain current I_D is recorded by a milliammeter (0–10 mA), and the gate-

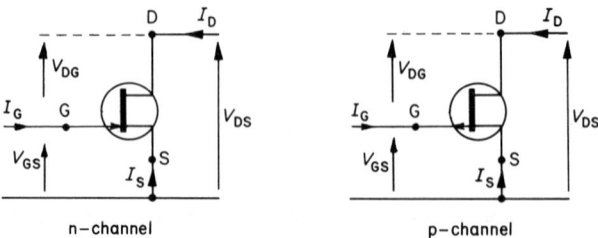

Figure 4.30. Polarity conventions in relation to the operation of n-channel and p-channel FETs

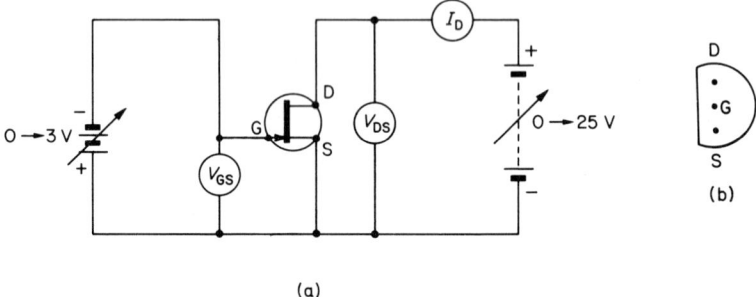

Figure 4.31. Determination of the drain characteristics of an n-channel FET

source voltage V_{GS} can be set and recorded at 0, -1, -2 and -3 V. An n-channel FET of type Semitron C94 or a 2N3819 is used. The base connections to the 2N3819 are shown in Figure 4.31(b).

The drain characteristics (I_D against V_{DS} for various values of the gate-source voltage) shown in Figure 4.32 exhibit two distinct regions of particular interest:

(i) *Below pinch-off*: in this region the drain current I_D increases with the drain-source voltage V_{DS}. The device acts like a variable resistance of which the magnitude is controlled by the gate-source voltage V_{GS}. Some types of FET are often used as voltage-controlled resistors.

(ii) *The pinched-off region*: the drain current I_D is virtually independent of the drain-source voltage V_{DS}. It is in this region that the FET is normally operated when used as an amplifier.

The *mutual characteristics* (I_D against V_{GS} for a given value of V_{DS}) can also be plotted from the results obtained from the circuit of Figure 4.31. A typical mutual characteristic for an n-channel FET is shown in Figure 4.33. Two terms of importance in connection with the FET are described below.

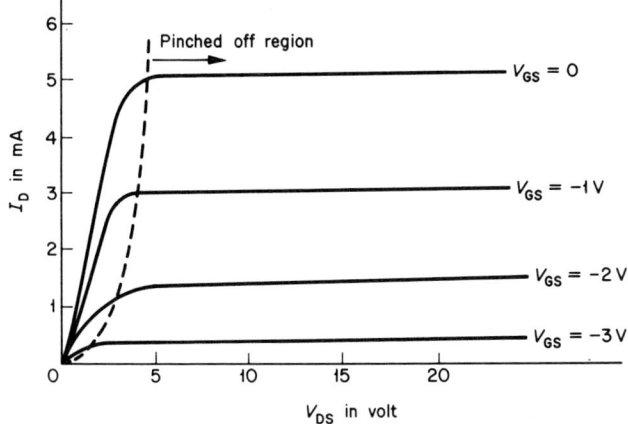

Figure 4.32. FET drain characteristics

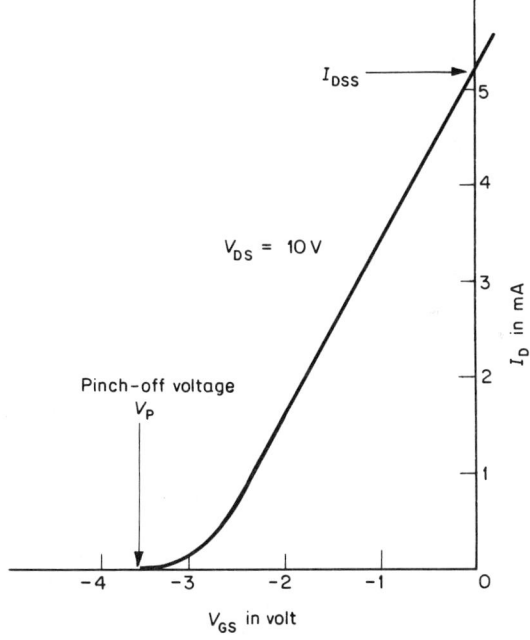

Figure 4.33. The mutual characteristic of an n-channel FET

(a) *The mutual conductance* (g_m) is defined as the change of the drain current for a change of the gate-source voltage. This is clearly related to the gradient of the characteristic of the drain current I_D plotted against the gate-source voltage V_{GS}, which is the mutual characteristic (Figure 4.33). The gradient of this characteristic is not constant, i.e. the characteristic is not linear. Hence, it is important to specify the change of I_D brought about by a *small* change of V_{GS}: Thus, in symbols, the mutual conductance g_m is given by

$$g_m = \Delta I_D / \Delta V_{GS} \qquad (4.12)$$

where Δ has its usual meaning as 'a small change of '. This is a conductance because it is the reciprocal of a resistance (which is voltage divided by current), hence the idea of a *mutual conductance*, sometimes called a *transconductance*.

The mutual conductance is also determined by the value of the drain-source voltage V_{DS}. It is clearly, therefore, a question of the mean value of V_{GS} about which the small change of V_{GS} takes place at a given value of V_{DS}. For $V_{GS} = -0.5$ and $V_{DS} = 10$ V, the mutual characteristic shown in Figure 4.33 is approximately linear with a slope of $\Delta I_D / \Delta V_{GS} = g_m = 2$ mA per volt, i.e. 2 mA/V.

(*b*) The *drain resistance* (r_d) is defined via the drain conductance (g_d) which is the change of drain current brought about by a small change of drain-source voltage at a given value of the gate-source voltage. Thus,

$$g_d = \Delta I_D / \Delta V_{DS}$$

Usually quoted is the reciprocal of this, i.e. $\Delta V_{DS} / \Delta I_D$ which is the drain resistance (r_d), at a specific value of V_{GS} and mean value of V_{DS}.

4.18 The Simple FET Amplifier

The simplest form of the FET amplifier is the so-called *common-source* type (Figure 4.34), which is based on an n-channel FET. In this a source resistor R_S is connected between the source and the common line (usually earthed) and has a value decided by the steady bias voltage needed on the gate. The constant current region of the drain characteristic of the FET (Figure 4.32) is the important part in the attempt to achieve a linearly operating a.c. amplifier, an amplifier for which the output voltage is an undistorted magnified replica of the input signal voltage. In this constant current region, the drain resistance r_d is very large and the effect on the drain current of any change in the drain-source voltage is negligibly small. Indeed, in this region, it is only the changes of the gate-source voltage — which means the input signal — that are effective. In fact, a useful good simple approximation is to put

$$\Delta I_D = g_m \Delta V_{GS}$$

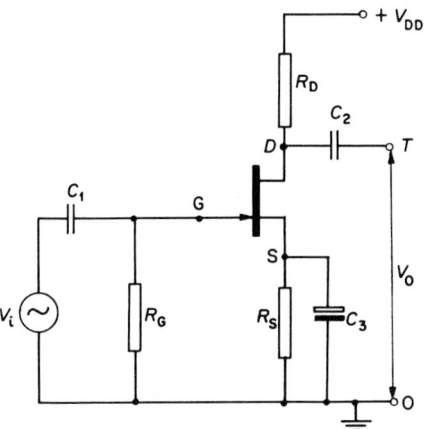

Figure 4.34. A common-source FET amplifier, based on an n-channel FET*

where $\Delta V_{GS} = v_i$, the input signal voltage. The voltage gain is then

$$A_v = v_o/v_i = v_o/\Delta V_{GS}$$

where v_o is the output voltage, which is given by

$$v_o = \Delta I_D R_D$$

so that

$$A_v = \Delta I_D R_D / \Delta V_{GS} = -g_m R_D \qquad (4.13)$$

assuming that $R_D < r_d$, where R_D is the value of the output load resistance in the drain circuit of the FET (Figure 4.34).

The steady bias voltage is decided by $I_D R_S$. The source is seen to be at a positive steady bias with respect to earth, taken to be at zero voltage. The input sinusoidal voltage signal v_i is at a mean steady value of zero, because it is with respect to earth. So the effective negative bias on the gate G is equal in magnitude to the positive bias on the source S, and hence

$$V_{GS} = -I_D R_S$$

The input capacitor C_1 allows the alternating input signal at its lowest frequency to pass, but prevents steady voltage from the source from affecting the biasing of the FET determined by $I_D R_S$.

---

* Note to Figure 4.34. Typical component values are $C_1 = 0.01\ \mu F$; $R_G = 10\ M\Omega$; $R_S = 390\ \Omega$; $R_D = 2\ k\Omega$; $C_2 = 0.01\ \mu F$ and $C_3 = 100\ \mu F$, with $V_{DD} = +15\ V$. Note C_3 is an electrolytic capacitor and there is an appropriate symbol for this.

The capacitor C_3 is a by-pass capacitor (usually an electrolytic capacitor of large capacitance to ensure that there is no alternating voltage produced across R_S (cf. Figure 4.13). The function of the capacitor C_2 is the same as in Figure 4.11.

An outstanding advantage of the FET over the bipolar transistor is that, due to its high input resistance, the input circuit of the FET (in an amplifier, for example) is virtually isolated from its output circuit. This greatly simplifies the theory involved in describing the circuit action.

Example 4.18

(a) *An n-channel FET is used in a simple common-source amplifier as in the circuit of Figure 4.34. Assuming that the drain resistance of the FET is much greater than the value of the load resistance R_D inserted in the drain circuit, that capacitors of sufficiently large values are used and that the mutual conductance of the FET is $g_m = 1$ mA/V, calculate the voltage gain in decibels provided by this amplifier, when $R_D = 10$ kΩ.*

From equation (4.13)

$$A_v = -g_m R_D = -10^{-3} \times 10^4 = -10$$

From Table 4.1 of section 4.13 it is seen that such a numerical value of the voltage ratio is equivalent to 20 dB.

(b) *In a single-stage FET amplifier for alternating voltage inputs (as in Figure 4.34) in a particular case the best value of the source resistor is 390 Ω. Given that the by-pass capacitor connected across R_S has a capacitance of 25 μF, calculate the frequency at which the reactance of this capacitance is 10 per cent of the value of R_S. Comment on the value of the frequency calculated.*

The reactance X_{C3} of C_3 is $1/2\pi f C_3$: substituting $C_3 = 25$ μF $= 25 \times 10^{-6}$ F gives

$$X_{C3} = 10^6/50\pi f \text{ ohm}$$

This is to be equal to 10 per cent of R_S at a frequency f given by $10^6/50\pi f = 39$. Therefore

$$f = 10^6/1950\pi = 163 \text{ Hz}$$

Comment: the by-pass capacitor across the source resistor is satisfactory provided that the alternating input frequency is not less than 163 Hz. For lower frequencies, a larger value of C_3 is required. To include the lowest frequency in the audio-frequency range, taken to be 20 Hz, it would be better to choose a value of $C_3 = 200$ μF, or even more.

4.19 Special Features of FETs

(a) The operation depends on the motion of majority carriers (electrons for an n-channel FET) only. This unipolar behaviour is less noisy; spurious signals (noise) occurring in an amplifier are often created by the recombination of electrons and holes in bipolar transistors.

(b) The input resistance is very high: typically several tens of megohm.

(c) They are relatively immune from nuclear radiations and consequently are often used in preamplifiers in radiation detectors.

(d) For an n-channel FET, the channel current is decreased when the negative voltage applied to the gate with respect to the source is increased, and it is fully conducting when the gate-source voltage is zero.

(e) Unlike the bipolar transistor, the FET is thermally stable. Any increase in temperature reduces the mobility of the carriers in the channel. As a result the drain current falls and the heating effect is reduced.

4.20 A Simple Constant-Current Circuit Based on an FET

The drain current I_D of an FET alters only insignificantly with change of the drain-source voltage V_{DS} (Figure 4.32) provided that the FET is operated in the pinched-off region, which means that V_{DS} is large enough for a given value of V_{GS}, the gate-source bias voltage.

This property of the FET enables a very effective and simple circuit to be built to provide a constant current through a load resistance R_L (Figure 4.35(a)). To establish the necessary bias on the gate relative to the source, the source resistor R_S is used to maintain the FET source at a positive potential with respect to the gate, i.e. the bias on the gate

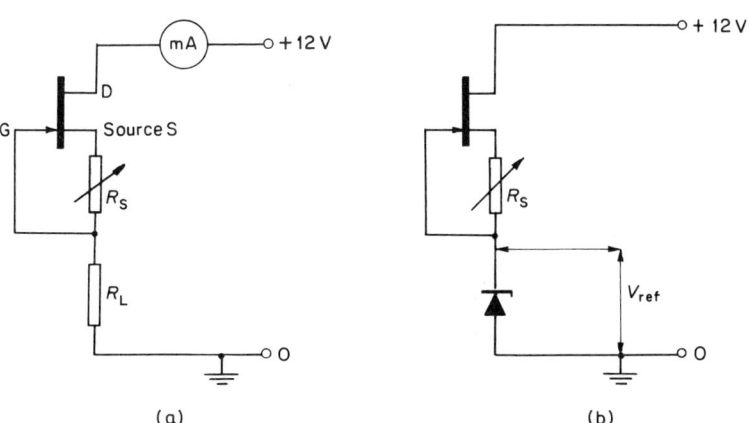

(a)   (b)

Figure 4.35. (a) A simple constant current circuit based on a FET and (b) use with a Zener diode to provide a high-quality reference voltage V_{ref}

relative to the source is negative at, say, $V_{GS} = -1$ V. In this circuit note that 100 per cent negative feedback is used (cf. section 4.10). If the drain current tends to increase slightly, the gate is made more negative with respect to the FET source, so this drain current (which flows through R_L) is reduced to its original value.

The milliammeter (mA) can be replaced by a high-quality resistor connected in series with R_1 and the potential drop across this resistor measured with a digital voltmeter (DVM) to determine the constant current flowing. This circuit is particularly valuable when a Zener diode (section 3.14) is used as a high-quality reference voltage. A Zener diode of approximately 5.6 V is selected (temperature coefficient is extremely small) and a constant current from the FET circuit is passed through the Zener as shown in Figure 4.35(b).

The constant current might be set at, say, 5 mA, and the circuit used must draw negligible current from the reference. The Zener diode can then be used to replace a standard cell. The standard cell or the high-quality reference voltage from the Zener can be connected in a potentiometer or to the non-inverting terminal of a differential operational amplifier (see Chapter 7).

4.21 The Metal Oxide Semiconductor Field Effect Transistor

Known as the MOSFET for short, this device is newer than the FET described above and is becoming very widely used. This is because it serves excellently in logic circuits for digital computation. MOSFETs have also enabled very large-scale integrated circuits (VLSI) to be made. A modern development from such VLSI is the microprocessor.

There are two main types of MOSFET: the n-channel and the p-channel. Each of these exists in two different versions: the enhancement (or 'normally-off') MOSFET and the depletion ('normally-on') MOSFET. In the main the interest is in the n-channel types of both versions. Sometimes a MOSFET is called an insulated gate FET (IGFET). The reason for this becomes clear on considering that the gate electrode G is separated from the silicon substrate by a layer of silicon dioxide, which is an excellent insulator. The normal FET (described in sections 4.15 and 4.16, with some of its applications in sections 4.16 to 4.20) is called a junction-gate FET (JGFET) to distinguish it from the MOSFET. The JGFET is also a 'normally-on' device. In the MOSFET, an immediate advantage of the silicon dioxide film (which is only about 100 nm thick) is that it retains its very high input resistance (about 10^6 MΩ, and some 100 times that of a JGFET) irrespective of the polarity of the voltage V_{GS} applied between the gate G and the source S. The very thin silicon dioxide film of a MOSFET needs to be protected against excessive voltages which might well occur when a very small charge is inadvertently placed on the gate electrode. With a capacitance of only perhaps 2 pF, a very small charge could be placed on the gate simply by touching it. A charge of only 1 microcoulomb (1 μC) would create a voltage given by charge/capacitance ($V = Q/C$, where Q is the charge and C is the capacitance) of 500 000 volts. The usual practice is to protect the gate of a MOSFET by means of a Zener diode incorporated in its structure between the gate and the substrate. Such Zeners are included in all integrated circuits based on MOSFET. The voltage on the gate with respect to the source can then never exceed the Zener breakdown voltage. During transport the leads of the MOSFET to its three electrodes G, S and D should be joined together electrically. The integrated circuits involving MOSFET are usually carried with their pins in conducting foam.

Example 4.21

A single-stage alternating voltage amplifier based on a JGFET (as in Figure 4.34) is supplied with an alternating voltage input signal of sine-wave form of frequency f and r.m.s. value v_i. There is a capacitor C_1 and a resistor R_G across the input signal source. It is known from a.c. theory that the r.m.s. voltage v across R_G (and so between the gate G and earth) is given by

$$v = v_i R_G / \sqrt{R_G^2 + (1/2\pi f C_1)^2}$$

If $C_1 = 0.01 \ \mu F = 10^{-8} \ F$ and $R = 10 \ M\Omega = 10^7 \Omega$, calculate the frequency f at which v is 0.9 v_i. Comment on the value of f so found.

It is convenient to use the square of the equation given, which is

$$(v/v_i)^2 = R_G^2/(R_G^2 + 1/4\pi^2 f^2 C_1^2)$$

Substituting the values of R_G and C_1 given in ohm and farad respectively and putting $v/v_i = 0.9$

$$0.9^2 = 10^{14}/(10^{14} + 10^{16}/39.5 f^2)$$

$$0.81(10^{14} + 2.53 \times 10^{14}/f^2) = 10^{14}$$

$$0.81(1 + 2.53/f^2) = 1$$

$$f^2 = 2.53/0.19$$

$$f = 3.65 \ \text{Hz}$$

Comment: at a frequency of 3.65 Hz the input circuit is such that the alternating voltage across the gate and earth is 10 per cent less than the input voltage itself, assuming that the input resistance of the JGFET is so high compared with 10 MΩ as to be negligible in its effect. At a frequency greater than 3.65 Hz, the fraction of v_i which appears at the gate is greater than 0.9. For audio-frequency operation over the range from 20 Hz to 20 000 Hz this is eminently satisfactory.

Exercise 4

1. Explain, with appropriate diagrams, how you would investigate the static characteristics of a p-n-p junction transistor in (a) common-base; (b) common-emitter connection.
   Sketch the curves which are obtained and comment on their special features. (A.E.B.)

2. Define the hybrid parameters of a transistor in common-emitter connection. How would you determine their values for an n-p-n silicon transistor?

3. Use a load-line to explain the terms *saturated* (i.e. *bottomed*) and *cut-off* for the a.c. operation of an n-p-n transistor in common-emitter connection.

4. Define the decibel and outline the advantages of using a logarithmic scale to express the power, voltage and current gain of an amplifier.

5. Assuming the load resistance in a common-emitter transistor amplifier to be small compared with the output resistance, draw a simple equivalent circuit for the amplifier. Explain how you could use this circuit to calculate the voltage gain, the current gain and the power gain of the amplifier.

6. Draw a circuit diagram of a common-emitter amplifier with resistive loading and a potential divider type of bias. Explain the action of the various components and the advantages of this type of bias.

7. Outline the methods of biasing a transistor and explain the advantages and disadvantages of each.

8. What is mean by (a) *positive feedback* and (b) *negative feedback*? In what type of circuit and for what purpose would each be used?
   In a voltage amplifier of open-loop gain A a fraction of the output voltage is fed back to the input in antiphase with the input signal. Derive an expression for the gain of this feedback amplifier and use this expression to discuss the merits of negative feedback.

9. What special features would you associate with a constant current source?
   Draw a circuit diagram and explain the action of the components in a source capable of passing a constant current of about 5 mA through a thermistor of which the resistance varies between 10 Ω and 2000 Ω.

10. Explain the principles involved in providing a stabilized voltage supply by the use of:
    (a) a Zener diode, and
    (b) a series control transistor.

11. Draw the circuit diagram and explain the operation of a transistor oscillator which provides a sinusoidal output. How would you measure the frequency of the oscillator?

12. An n-p-n junction transistor used as a voltage amplifier in common-emitter connection has hybrid parameters $h_{ie} = 1$ kΩ, $h_{oe} = 5 \times 10^{-5}$ siemen and $h_{fe} = 50$. The load resistance is 2 kΩ and the input e.m.f. is 20 mV. Draw an equivalent circuit for this amplifier and calculate the output voltage and output power.

13. 'The principle of operation of a field effect transistor is quite different from that of a bipolar junction transistor'.
    Provide evidence in support of this statement.

14. With the aid of an appropriate circuit diagram explain how the drain characteristics of an n-channel field effect transistor are determined. Sketch the drain characteristics and comment on the prepinch-off and pinched-off regions.

15. Draw the circuit diagram of a single-stage alternating voltage amplifier based on a JGFET. A capacitor C is joined between one of the input terminals (the other input terminal is earthed) and the gate G of the JGFET. Between G and earth is joined a resistor R_G of value 10 MΩ.

Calculate the value of C which is needed to ensure that the alternating voltage between the gate and earth is not more than 10 per cent less than the input voltage at a frequency of 20 Hz.

5 Some other semiconductor devices and basic applications

There are now a number of semiconductor devices which are very useful, but they are not formal transistors or formal diodes. The number of them is forever increasing: the ones considered here are the photoelectric devices, the thermistors, the light emitting diodes (LEDs) and the liquid crystal displays (LCDs).

5.1 Photoelectric Effects in Semiconductors

If photons of energy hv are incident on the depletion region of a surface type p-n junction electron-hole pairs are created. The electric field existing across the depletion region causes motion of the free charges created. Equilibrium may be restored by current flow in an external circuit.

The energy gap for silicon is 1.1 eV (section 1.16). Photons with energies exceeding this value are therefore effective. The corresponding threshold wavelength λ_t of the incident radiation is hence given by

$$hv = hc/\lambda_t = 1.1 \text{ eV}$$

In this relationship h must be in eV second instead of the usual joule second. As $h = 6.625 \times 10^{-34}$ joule second and $1 \text{ eV} = 1.6 \times 10^{-19}$ joule, so

$$h = \frac{6.625 \times 10^{-34}}{1.6 \times 10^{-19}} = 4.13 \times 10^{-15} \text{ eV s}$$

Putting $c = 3 \times 10^8$ m/s,

$$\lambda_t = \frac{4.13 \times 10^{-15} \times 3 \times 10^8}{1.1} = 1.13 \times 10^{-6} \text{ m} = 1130 \text{ nm}$$

The visible region of the spectrum extends in wavelength from 7×10^{-7} m (700 nm) to 4×10^{-7} m (400 nm) approximately. The silicon detector therefore gives a photoelectric effect with visible light and infra-red radiation up to wavelengths of 1100 nm approximately.

130

The photoelectric effect in this case is more specifically a *photo-conductive effect* in that the incident radiation on the p-n junction creates current carriers.

A number of factors at present limit the efficiency of these photo-electric semiconductor cells to about 10 per cent, i.e. the electrical power output is about 10 per cent of the energy per second (power) due to the incident radiation. These factors include:

(a) The electron-hole pairs created recombine before they are collected by attached electrodes.

(b) The radiation (light) is partly reflected from the front surface of the cell, and so some of it does not traverse the depletion region about the p-n junction.

(c) The electrical resistance of the cell causes power loss because passage of current through it results in dissipation by heat.

Nevertheless, surface type p-n junction silicon cells have become important not only as photovoltaic detector devices for radiation but also as *solar batteries*.

In full sunlight, such a silicon cell will develop an e.m.f. of approximately 0.6 V on open circuit (i.e. no load resistance present). The larger spacecraft and satellites frequently contain a solar cell panel containing about 5000 such cells to provide power of 100 W at 30 V. These cells are connected in rows in parallel to reduce resistance and with these rows connected in series to provide the output voltage.

Among the other devices which exploit the creation of electron-hole pairs by incident photons are *photodiodes* and *phototransistors*. The structure of a photodiode is shown in Figure 5.1(a). The diode is always operated with reverse bias, with up to 30 V across the p-n junction. The dark resistance (resistance with no incident light) at this maximum voltage is about 20 MΩ. The corresponding dark current is then $30 \text{ A}/(2 \times 10^7) = 1.5 \ \mu\text{A}$. This current is obviously due to minority carriers.

A converging lens is normally used to focus the incident light at the photodiode in the otherwise light-tight capsule. This light therefore impinges in the immediate area of the p-n junction behind the lens window. Alternatively, a separate, external lens and a plane window may be used. Electron-hole pairs are therefore created near the surface to form the reverse current which is now large compared with the dark current. This reverse current is a linear function of the illumination produced by the incident light. It might be limited to say, 3.0 mA to avoid temperature rise due to heating effects at the junction. The device can operate up to a frequency of 50 kHz, and thus can easily detect the fluctuations in light intensity from discharge tube lighting.

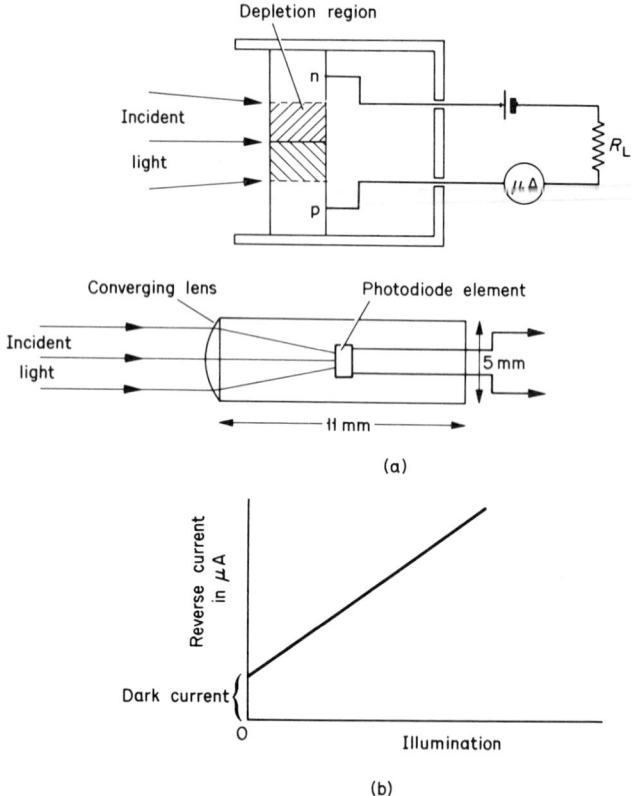

Figure 5.1. The photodiode

Operating in the manner described, the photodiode behaves as a current generator with a high internal resistance (Figure 5.1(b)). A resistance in the external operating circuit has little effect on the current flowing. The voltage developed across this external circuit has a magnitude dependent upon the illumination by the incident light. This voltage can be used to operate a circuit or device in light-controlled apparatus. An oscillograph connected across this resistance will display the waveform of any fluctuations in the source of light that produces the illumination of the photodiode.

The germanium photodiode shown in Figure 5.1(a) has a peak response to radiation of wavelength 1500 nm in the infra-red but still responds well in the visible region.

A small filament lamp operated from a constant voltage transformer or a constant current source is ideal for experiments with photodiodes.

The *phototransistor* is simply a junction transistor in the capsule of which is a region through which light may be passed to impinge on the sensitive base region. The current carriers produced on incidence of light on the base region are amplified by the transistor action. The electron-hole pairs produced by the light are equivalent to base current; the number produced therefore controls the much larger current which flows between the emitter and collector.

A *photoconductive cell* is also formed by the production of a thin film of semiconducting material between two metal electrodes (Figure 5.2).

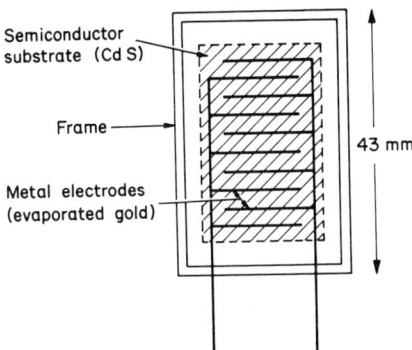

Semiconductor substrate (Cd S)

Frame

Metal electrodes (evaporated gold)

43 mm

Figure 5.2. A photoconductive cell

The semiconductor material used is cadmium suphide (CdS) or lead sulphide (PbS) or indium antimonide (InSb) in the form of a thin polycrystalline film. Incident light traverses the thin semiconductor film and the energetic photons create in it electron-hole pairs which increase its conductivity.

The circuit used simply involves the photoconductive cell in series with a microammeter or milliammeter and a steady voltage supply. As the cell resistance changes depending on the light illumination, the current recorded by the series meter changes. For visible light, cadmium sulphide is mainly used. For infrared radiation, lead sulphide cells are effective.

The dark resistance of a typical CdS photoconductive cell is approximately 100 MΩ. The maximum voltage that can be applied across it is 300 V. The current through the cell is limited by a series resistor to prevent undue power from being dissipated in the cell. The dissipated maximum power is typically 0.5 W.

When the illumination is removed, the current decays exponentially over a period of about one minute.

A widely used application of cadmium sulphide photoconductive cells is in exposure meters used in photography. The operating e.m.f.

is a few volts. They are sufficiently sensitive to be able to record the exposure required even when the scene to be photographed is illuminated only by bright moonlight.

5.2 Thermistors

Thermistors are semiconducting resistors with a large negative temperature coefficient of resistance. Thus, as the temperature is increased, the resistance decreases. The temperature change can be caused either in the surroundings in which the thermistor is immersed or by heat generated within its element due to the passage of current through it.

Over a specified temperature range, the temperature coefficient of resistance, α, is defined as

$$\alpha = \frac{\text{rate of change of resistance with temperature}}{\text{original resistance}}$$

which, in mathematical terms, becomes

$$\alpha = \frac{1}{R_T}\frac{dR_T}{dT} \tag{5.1}$$

where R_T is the resistance at a temperature T K. For thermistors at room temperature ($20°C = 293$ K), α has a value of -0.06 per degree Kelvin (-0.06 K^{-1}) which is -6 per cent K^{-1}. For metals, α is positive and about $+0.003$ K^{-1}, i.e. 0.3 per cent K^{-1}. The thermistor is clearly, therefore, much more sensitive to temperature change than a metallic resistance (as used, for example, in the normal resistance thermometer).

Thermistors are one of the few semiconducting components manufactured which are not made from single crystals. Although a slice of intrinsic germanium has a useful resistance against temperature characteristic (Figure 1.13) much research has enabled reliable and cheap components to be made from a mixture of semiconducting oxides, commonly Fe_3O_4 and $MgCr_2O_4$.

The resistance of a commercial thermistor is normally specified to be within ± 20 per cent of a stated value at a given temperature. On occasions the resistance against temperature characteristic of a component may not be accurately reproducible.

A plastic binder added in manufacture to the carefully prepared oxide mixture enables the material to be extruded into rods, pressed into discs or formed into beads between two platinum wires. The component so formed is then sintered at high temperature which causes the material to shrink on to the wires and make good electrical contact. Miniature bead thermistors are often glass-encapsulated or enamelled for protection. A number of commercially available

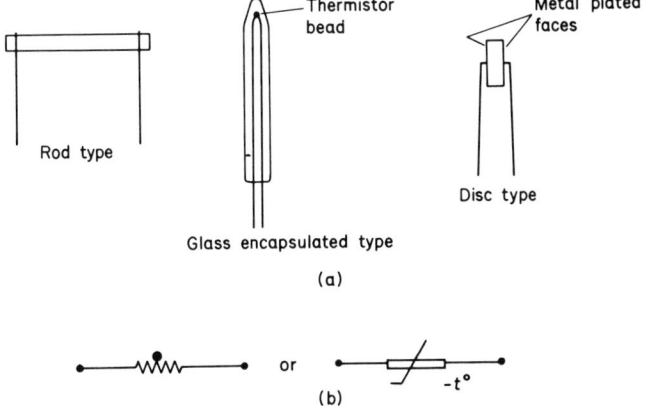

Figure 5.3. (a) Thermistors and (b) circuit symbol

thermistors is shown in Figure 5.3 together with the conventional circuit symbol for a negative temperature coefficient device.

The relation between the resistance R and temperature T for a thermistor (Figure 5.4(a)) can be expressed in the form

$$R_T = A \exp(B/T) \tag{5.2}$$

where R_T is the resistance at T K and A and B are constants for a particular component, A being in ohm and B in degree Kelvin. Hence

$$\frac{dR_T}{dT} = A\frac{d}{dT}\left[\exp\left(\frac{B}{T}\right)\right]$$

$$= -A\left[\frac{B}{T^2}\right]\exp\left[\frac{B}{T}\right]$$

$$= \frac{-AB \exp(B/T)}{T^2} = \frac{-BR_T}{T^2}$$

Hence

$$\alpha = \frac{1}{R_T}\left(\frac{dR_T}{dT}\right) = \frac{-B}{T^2} \tag{5.3}$$

The constant B depends on the material. As α is specified as per degree Kelvin $(1/T)$, it follows that B must be in degree Kelvin. Usual values of B are between 2000 and 6000 K.

Figure 5.4(b) shows the voltage against current characteristic of a thermistor being heated by the current passing through it.

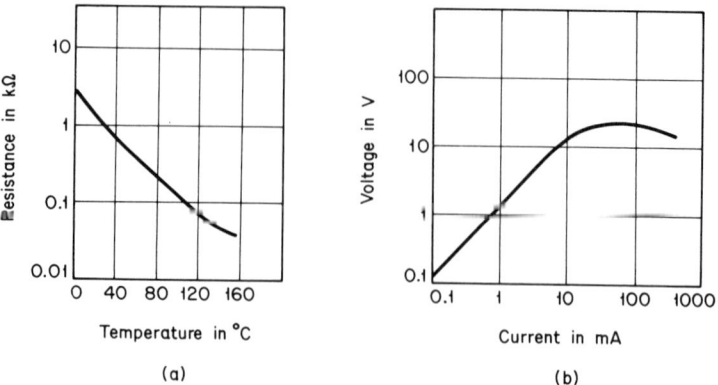

Figure 5.4. (a) Resistance against temperature characteristic of a thermistor and (b) voltage against current characteristic of a thermistor

Example 5.2

A negative temperature coefficient bead thermistor located in the end of a glass rod has a resistance of 2000 Ω at 300 K and a value of B in equation (5.3) equal to 2600 K. Calculate its resistance at a temperature of 500 K.

For a negative temperature coefficient thermistor, the resistance temperature relationship is given by equation (5.2):

$$R_T = A \exp(B/T)$$

At $T = 300$ K, substituting $R = 2000$ Ω and $B = 2600$ K gives

$$2000 = A \exp(2600/300) = A \exp(8.67)$$

At $T = 500$ K, the resistance is R_{500} given by

$$R_{500} = A \exp(2600/500) = A \exp(5.2)$$

Hence

$$\frac{R_{500}}{2000} = \frac{\exp(5.2)}{\exp(8.67)} = \exp(5.2 - 8.67)$$

$$\exp(-3.47) = 0.0311$$

$$R_{500} = 2000 \times 0.0311 = 62 \ \Omega$$

5.3 Some Applications of Thermistors

The most obvious application for a thermistor is to the measurement of temperature. Any one of the three circuits shown in Figure 5.5 can be used; the current through the thermistor should be kept as small as possible. The bridge-type circuit (Figure 5.5(c)) is very sensitive and also compensates for changes in the ambient temperature.

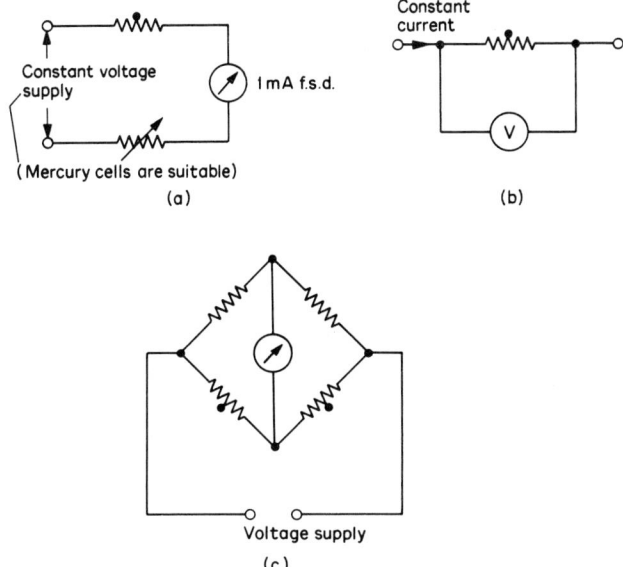

Figure 5.5. Thermistors in temperature measurement

The following advantages result from the use of thermistors for temperature measurement:

(a) The thermistor has a useful temperature range: −70°C to 300°C.

(b) It has high sensitivity: −6 per cent change in resistance per degree Celsius change in temperature at room temperature.

(c) Because of the high resistance of the thermistor, the length of the connecting leads and their change of resistance with temperature is generally of no consequence.

(d) A small thermistor bead has a very small heat capacity as a temperature-sensing device. It is consequently able to respond to rapid changes of temperature.

(e) The element is more robust than the thermocouple and the platinum resistance thermometer.

Two disadvantages of thermistors are:

(i) They are not used for the high-precision measurement of temperature because their characteristics are not perfectly reproducible like those of a platinum resistance thermometer or a thermocouple.

(ii) Their characteristics exhibit hysteresis which depends on the previous history of usage of the thermistor.

A thermistor thermometer has to be calibrated; in a simple experiment this is done against a mercury-in-glass thermometer.

Thermistors are frequently used in conjunction with relays. For example, a sensitive fire-alarm circuit has a thermistor in series with the energizing coil of the relay. When the thermistor is at room temperature its resistance is large enough to restrict the current to below the value at which the relay is energized. Should the temperature of the surroundings rise, the resistance of the thermistor will fall, the relay becomes energized and the current to the fire-alarm bell (which is much larger than the current in the relay energizing coil) is switched on.

Thermistors with *positive* temperature coefficients of resistance are available but the range of resistance values is severely limited. They are made of barium titanate.

Below a certain critical temperature, the resistance of such a thermistor is almost constant. Once the critical temperature is exceeded, the resistance rises rapidly with temperature increase. By controlling the process of manufacture of this type of thermistor, the critical temperature can be arranged to be, for example, 100°C, 115°C or 130°C (Figure 5.6(a)).

A positive temperature coefficient thermistor has a circuit symbol similar to that of the negative coefficient device except that an open

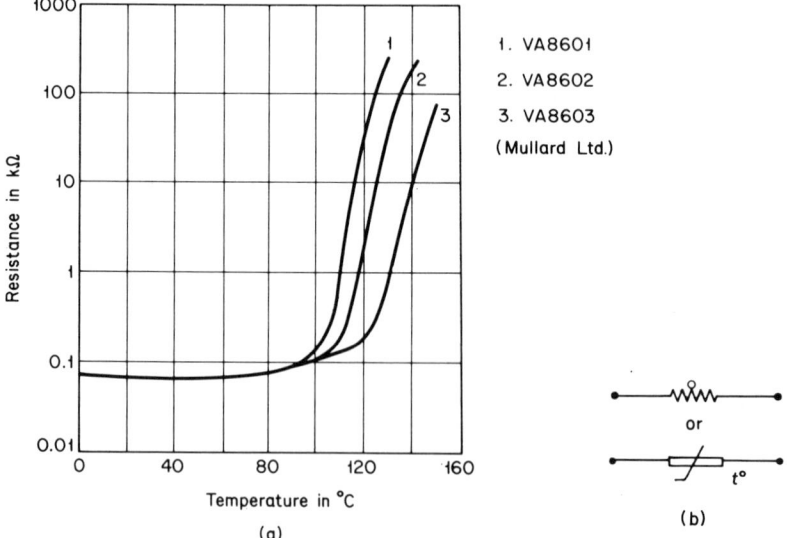

1. VA8601
2. VA8602
3. VA8603
(Mullard Ltd.)

Figure 5.6. (a) Characteristics of positive temperature coefficient thermistors and (b) circuit diagram symbol for a positive coefficient thermistor

circle on the resistor symbol is used in place of the black dot (Figure 5.6(b)).

5.4 Light-Emitting Diodes (LEDs)

Whereas the photodiode is a semiconductor device which produces an electric current when light is incident upon it, the light-emitting diode (LED) is the opposite: it is a semiconductor p-n junction diode which emits light when a current is passed through it.

The p.d. across the LED is necessarily in the forward direction to cause it to emit light; this means that the p-type side is made positive with respect to the n-type side. The electrons which traverse the junction from the electron-rich n-type semiconductor material reach the p-type material and there combine with the plentiful positive holes. On combination, the electrons lose energy; in doing so, they cause the emission of radiation, which is in the visible region of the spectrum if the energy they lose is somewhere in the range of energies of the photons of the visible spectrum between the violet and the red. To produce light which is readily seen by the human eye, which has a peak sensitivity in the green, at a wavelength of about 555 nm, the most suitable material for making LEDs appears to be nitrogen adulterated gallium arsenide — a III–V semiconductor (section 1.10).

The light output from an LED increases with the current passed through it on the application of the forward voltage. For example, a small LED can be operated with a forward voltage of 2 V to produce a current of about 10 mA, so that it gives enough light to be seen readily in a normally-lit room. Unfortunately, 10 mA is a fair amount of current and is a considerable drain on a small dry cell such as is used in hand electronic calculators and digital watches. At present, therefore, the liquid crystal display (LCD), described in section 5.5, is the type used almost exclusively in such digital devices. A resistance connected in series with a LED is essential to limit the forward current through the diode. With a 5 V stabilized supply, as used with TTL logic gates, a 150 Ω resistor is connected in series with each LED.

5.5 The Liquid Crystal Display (LCD)

For over 90 years it has been known that certain organic materials, when between their liquid and solid states, have a rod-like molecular structure. The molecules can be aligned by applying an electric field. This alters the optical state of the liquid crystal (as it is called); for example, it can be changed from being a good absorber (i.e. turbid) to a good reflector of light. As is now known, the liquid has a long-chain molecular structure and the effect of the applied field is to orientate the molecules to an extent depending on the strength of the electric field. The correct kind of such a liquid crystal is called 'nematic'. A liquid crystal display cell is illustrated schematically in Figure 5.7(a).

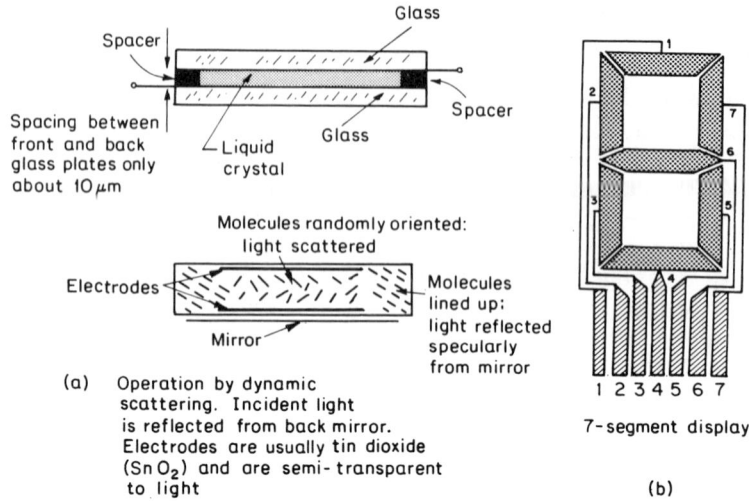

Figure 5.7. Liquid crystal displays

It comprises a liquid crystal in between two parallel glass plates which are about 10 μm apart. (1 μm $= 10^{-6}$ m). The surfaces (which eventually face one another in the cell) of the glass plates are first coated with a photosensitive oxide. By means of photolithographic methods (similar to those used in the fabrication of integrated circuits, as described in Chapter 6) the required display pattern together with the associated electrical leads and contacts are etched into the surface of the oxide layer on each plate. For example, the surface of one of the plates can thus be coated with stannous dioxide (which is a chemically deposited, light-transparent film which has good electrical conductivity) to produce the seven-segment display (which enables any of the digital numbers 0 to 9 to be formed) shown in Figure 5.7(b), whereas the opposite glass plate surface has a suitable stannous dioxide electrode on it to act as the common return for all segments. These electrodes reflect light to an extent dependent upon the electric field which is present in the liquid crystal and so on the p.d.s. applied to the electrodes.

It is readily shown that seven bars can be used to delineate any number between 0 and 9. In Figure 5.7(b), for example, is shown the bars (where all seven are illuminated) to give the number 8. An essential difference here between the use of LEDs and LCDs is that in the former, light is *emitted* by the bar; in the latter it is *reflected* to a greater or lesser extent. Thus, an LCD, unlike an LED, does not show up in the dark. However, a great advantage over the LED is that the LCD can be operated from a 1 to 3 V source of supply and requires a current of only about 3 μA. They are consequently ideal (in contrast with LEDs) for such devices as digital wrist-watches. However, direct current (d.c.) is not used in LCDs because of the deleterious electrolytic effect.

The liquid crystal effect required — which is due to the alignment of the long-chain molecules parallel to one another — occurs over a specific temperature range, typically 5°C to 60°C, which includes ambient temperatures over most of the world's surface. There are also liquid crystal devices of a somewhat different kind which are markedly temperature-sensitive and can be used to indicate specific temperatures.

Exercise 5

1. The reverse current through a junction diode increases considerably when visible light falls on the junction. Why does this occur?
   (A.E.B., part)

2. The energy gap (E_g) in germanium is 0.75 eV. Calculate the maximum wavelength radiation which is capable of creating an electron-hole pair in germanium. (The Planck constant $h = 6.62 \times 10^{-34}$ J s; 1 eV $= 1.6 \times 10^{-19}$ J; velocity of light in free space, $c = 3 \times 10^8$ m s^{-1}.)

3. (a) Why are transistors mounted within air-tight and light-tight containers?
   (b) Why is silicon usually preferred to germanium as a transistor material?

4. Sketch a curve showing the variation of resistance with temperature for a negative temperature coefficient thermistor. What is the form of the equation that approximates to this curve?

5. For the construction of a device for sensing temperature changes, compare and contrast the use of (a) a metal and (b) a thermistor material where, in both cases, the variation of resistivity with temperature is determined.

6. What advantages and disadvantages result from the use of thermistors for temperature measurement?

7. Write an account of the light-emitting diode or LED. In your account, described in outline:
   (a) the principle of operation;
   (b) the advantages and disadvantages.

8. With the aid of a suitably labelled diagram describe a liquid crystal display device.
   What is the main advantage of the LCD as compared with the LED? Show, with a suitable diagram, how any integral number from 0 to 9 can be delineated by the use of seven straight segments.

6 The fabrication of integrated circuits

6.1 Active and Passive Components

An active component is conveniently defined as one of which the electrical behaviour can be altered by the input voltage or current (i.e. signal) applied to it: the chief example is the transistor. A passive component is one of which the electrical behaviour is not changed in response to an external signal: the examples are the resistor, the capacitor and the inductor. The last of these cannot be made of dimensions small enough to be constructed on a silicon chip so inductors are not used in integrated circuits. For present purposes, the diode is best classified as an asymmetrical component: for an applied p.d. of fixed polarity, it offers a ready path to the flow of current in one direction but not in the opposite direction. The diode is a non-ohmic device (or non-linear in the sense that the current I through it due to the application across it of a voltage V is *not* such that I is directly proportional to V).

6.2 Why were Integrated Circuits Developed?

There are far more electronic components in the automatic digital computer than in any other single piece of apparatus. As explained more fully in Chapter 8, the heart of all kinds of digital computer is the very fast switching circuit. This must be relatively simple in structure but of high reliability. In the 1950s, when large computing systems were first being developed, these switching functions were performed by circuits based on thermionic vacuum tubes (radio valves). These were mounted on circuit boards and wired up by soldering connecting wires to the necessary passive components. Each such circuit board was simple in structure but thousands of them were required to do the calculations and subsequently store or print out the results. The first of the large digital computers of the electronic type was developed in 1946 by Eckert and Mauchly at the Moore School of Engineering in Pennsylvania, USA. It contained 18 000 valves! Imagine the very large room needed for such a machine, the fans required to dissipate the heat developed, the air-conditioning needed to keep the ambient conditions fairly constant and the army of maintenance engineers moving from one bank of switching circuits to the next, replacing faulty valves and re-making faulty connections. More time was spent in maintaining the equipment than actually using it. When the transistor was used to replace the radio valve, the size of the computer was reduced, the power requirements fell and the reliability was much improved.

The outstanding breakthrough, leading to microelectronics — the twentieth-century revolution — came, however, when the transistors, the resistors, the capacitors and their interconnections were all fabricated on the same chip of silicon, which is the integrated circuit. This advance enjoyed enormous financing in the USA because of the demands in space vehicles for computers (used for flight control) and other electronic devices to be as small and light as possible.

Because of the fundamental importance of the NAND gate (see section 8.9) in digital electronics, integrated circuits are usually classified in terms of the effective number of NAND gates on the chip. This leads to the following classification:

Small-scale integration (SSI) relates to circuits with between 1 and 12 gates on a single chip.
Medium-scale integration (MSI) is for circuits with between 13 and 100 gates on a single chip.
Large-scale integration (LSI) is for 100 to 500 gates.
Very large-scale integration (VLSI) is for more than 500 gates.

Development continues to fabricate ever more switching circuits or gates on a single chip. Not only does the number of gates determine the size of the memory used in the digital computer (see Chapter 8), but also, as the number of gates per unit area of the chip is increased, the effective cost per gate can be reduced.

6.3 Integrated Circuits

Since about 1960 there has taken place this fantastic development in the integrated circuit (i.c.). An i.c. may contain several tens of thousands of active and passive components all within the surface (essentially) of a chip of silicon of side length not more than 6 mm and thickness about 0.5 mm. The chip is broken in a pre-determined way from a disc or wafer of silicon of thickness 0.5 mm and diameter 50 or 100 mm.

Integrated circuit technology is usually based on either bipolar junction transistors or on MOSFETs. In either case, the circuit concerned has to include, as well as the active components, the passive ones and the necessary connecting leads together with the terminals (lands). The terminals are for the input and output connections and the power supplies. The development of the integrated circuit has brought about great interest in microelectronics. Some of the more important applications of microelectronics are digital watches, pocked electronic calculators, computers, microprocessors, communication satellites and control circuits.

6.4 The Production of a p-n Junction by Diffusion

The most important method of producing the necessary p-n junctions in integrated circuits is by diffusion. This was first shown to be the case in 1958 by J. Hoerni of the Fairchild Semiconductor Corporation in the USA. The method has already been described in section 2.4, dealing with the principles of manufacture of a semiconductor planar diode. Diffusion methods make it possible to control very precisely the concentration and the concentration gradient of a dopant over a very small region of a semiconductor.

6.5 Some Aspects of the Fabrication of Integrated Circuits

The basic methods of integrated circuit manufacture were established by 1960 in the USA. The use of photolithography (or, more simply, masking methods and the use of photography) to form the surface configuration of planar transistors was an American development in the mid-1950s. The number of industrial concerns throughout the world which are engaged with the manufacture of integrated circuits is small. The intimate details of some of the methods they use tend to be guarded secrets so that only general statements of a wide applicability can be made here.

The manufacture of integrated circuits (at least 2000 different ones are available) involves means whereby on a chip of silicon of not more than 6 mm on the side, there is produced a number of active and passive components which are connected together in specified ways and, equally important, are electrically insulated from one another as necessary and from the substrate material, where required. There are two ways of providing electrical insulation where required: the first (and the more important) is to make use of the insulating properties of silicon dioxide; the second, is to introduce, where needed, p-n diodes as rectifiers connected with the applied p.d. the wrong way round, i.e. so that no conduction occurs.

Resistors are made in i.c.s in two different ways. For values up to about 5 kΩ, a thin (often meandered) ribbon of p-type silicon is produced in an island (extremely small) of n-type silicon and so that the p-type ribbon is always maintained at a negative potential with respect to the n-type (i.e. as a diode, this p-n junction is reverse biased). Alternatively, silicon dioxide insulation is used. The second way, used for larger resistance values, is to make use of the voltage- or current-controlled resistance of a transistor (see Chapter 4) which enables resistances up to values of a few 100 kΩ to be simulated.

In i.c.s. capacitors are seldom of values greater than a few picofarad (1 picofarad = pF = 10^{-12} F). This is because there is no way with i.c.s of interleaving dielectric films between plane-parallel conductors of significant areas, in the manufacture of capacitors as discrete components. However, to form a capacitor, a doped semiconductor is used to form one plate, silicon dioxide (SiO_2) is the dielectric and an aluminium top plate is deposited on the oxide to form the second plate. Use is also made of the capacitance of a p-n junction (see sections 3.3 and 4.1).

In the making of an i.c., the point of beginning is invariably the rod of diameter, say, 50 mm or 100 mm and length several cm of single-crystal silicon (section 2.2). This silicon is either pure or doped n-type or p-type. This cylindrical rod of silicon is first cut into slices by the use of a diamond saw where each slice (usually, and henceforth here, called a wafer) has a diameter of 50 to 100 mm and a thickness of about 0.6 mm. By the use of a diamond there is scribed on to the surface of this wafer several parallel straight lines at 6 mm intervals, first in one direction and then in the perpendicular direction. Therefore, the wafer, after the process of making the i.c.s is completed, can be readily broken up into a few hundred 'chips' each of side length not more than 6 mm and each of which contains (essentially on its surface) the whole of an integrated circuit which, in the case of VLSI, may be 10 000 components. Of these chips, with the i.c.s formed on them, only about 25 per cent pass the final, computer-operated, electrical circuit checking; the rest are rejected.

In all cases of making i.c.s the surface of the silicon wafer is first lapped and polished. This wafer usually then has its surface covered with a very thin expitaxial film of n-type or p-type silicon. In the epitaxial growth of a film from a vapour phase on the surface of a single crystal of the same element, this film takes up the same crystal-lattice orientations as in the single crystal. In several cases, the basic single crystal itself is of n-type or p-type silicon.

The main preliminary steps in the making of integrated circuits based on n-p-n bipolar transistors are illustrated by Figure 6.1. These steps are as follows (they should be studied in conjunction with Figure 6.1).

(i) The entire silicon wafer (which has a p-type epitaxial film on its surface) is heated to about 1000°C (the temperature has to be controlled to within 1°C) in a stream of an oxygen-bearing gas so that its surface becomes covered with a thin film of known thickness of silicon dioxide. Several wafers, usually a hundred or more, are processed together. The p-type surface layer of the

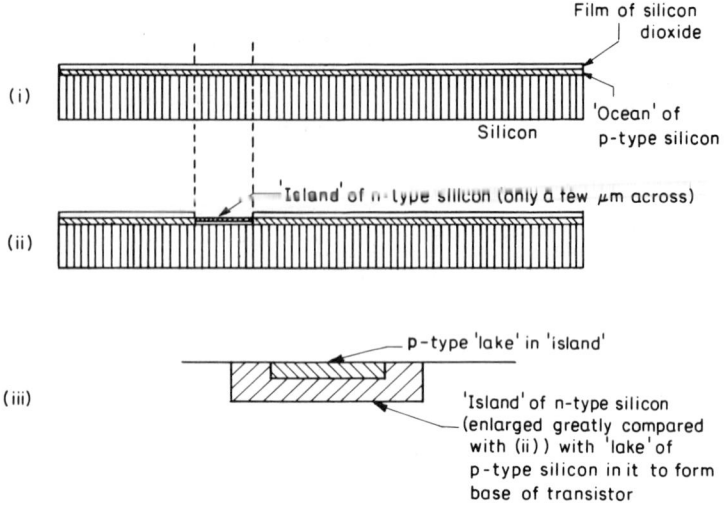

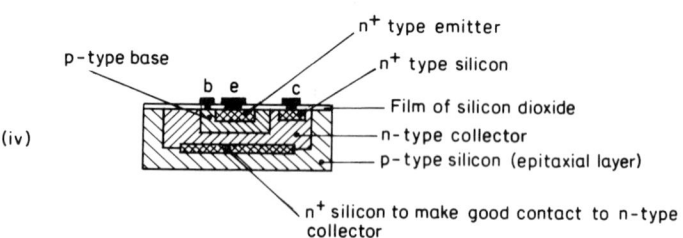

Figure 6.1. Some basic stages in the fabrication of integrated circuits (all dimensions are greatly exaggerated and out of scale; repetition of some of stages is needed in practice; n$^+$ type is heavily doped n-type silicon)

silicon (beneath the silicon dioxide film) is sometimes called an 'ocean' (even though it is only 50 to 100 mm across!).

(ii) In this ocean are created separate n-type silicon 'islands' (each only a few μm across). Such an 'island' is needed for each of the transistors and for each of the passive components in a given chip. These islands are all maintained at a positive potential with respect to the ocean of p-type silicon so as to be electrically isolated therefrom.

(iii) The n-p-n transistors need to have each an n-type emitter, a p-type base and an n-type collector. To form the bases of the transistors in each of the islands, a lake of p-type silicon is formed within each n-type island.

(iv) To form the emitters of the transistors, yet smaller islands of n-type silicon are produced in each of the lakes. These emitter islands are each about a quarter of the area of the collector islands.

In these stages of fabrication (and similar for MOSFET i.c.s although the stages — except for the surface oxidation — will then be different) it is essential to make use of some method(s) whereby silicon (n-type or p-type) can be introduced in specified very small areas of each of the chips. This means a *masking* technique involving a material which acts as a mask in that it contains apertures in specific places through which the correct kind of dopant for the silicon can be introduced. Fortunately, silicon dioxide films can be used as such masks — a very useful facility of this material which has had a great influence on the use of the readily oxidized silicon as *the* element for i.c. manufacture. This idea of using silicon dioxide as a masking material is really assumed in the outline description of the main preliminary stages in the fabrication of a silicon planar diode (section 2.4). In this description, in stage (*b*), reference is made to the fact that apertures are cut in the oxide layer to expose the silicon which subsequently, in stage (*c*), has boron diffused into it through the apertures in the silicon dioxide film. The same ideas are implicit in the fabrication of integrated circuits. The valuable masking properties of silicon dioxide films in this connection depend on the facts that the acceptor dopant element boron and the donor dopant phosphorus (but not arsenic) diffuse at much slower speeds at 1000°C through silicon dioxide than through silicon.

There remains the necessity to have made in the silicon dioxide film, on the surface of the silicon, apertures of various very small sizes and in locations each specified to within 1 μm so that the 'islands' and 'lakes' of extremely small dimensions can be formed (see items (ii), (iii) and (iv) above). This involves a different sort of masking and the use of what are preferably called 'working plates', and also methods which have been, in effect, borrowed from (with extraordinary refinements) the printing industry's processes of photolithography and photogravure.

In the preparation of the working plates, the first step is to design the integrated circuit from the points of view of its functions, component characteristics, voltages and currents and layout of active components, passive components, electrical connections and lands. This design is greatly assisted by the use of the digital computer both to simulate the circuit required and to display it graphically. In some cases, especially for LSI and VLSI, master drawings of aspects of the circuit are made. This practice is greatly assisted (or has been superseded) by storing in the memory of a digital computer numerical information about the sizes and locations of the parts of the components in an i.c. Consequently, this information is about the aperture sizes and locations which need to be made in the silicon dioxide film on the silicon wafer.

The computer, stored with these data, is used to control the scanning of a spot of light across a photographic plate to produce a *reticle*. The pattern in each reticle is linearly ten times the size of the i c to be produced. An image of this reticle is focused by optical lenses (of extremely high quality) on to the photomask and a very accurately made, specially designed mechanical system is used to cause this image to be reproduced hundreds of times so as to form the image pattern on each of the chips pre-diamond-scribed within the silicon wafer. The demands on this stage of the process can be imagined when it is realized that each integrated circuit component has to be registered in size and location to within 1 μm! From the photomask is made photographically a series of submasters; these are used, in turn, to produce the number of working plates employed in the actual fabrication process.

It is also clearly necessary to utilize some method whereby the silicon dioxide film is etched away in the desired locations to form the apertures set by the photomask. For this, a photoresist is used, which is coated as a uniform, thin, dried, adherent layer on top of the silicon dioxide film on the surface of the silicon. This photoresist is a plastic material which dissolves readily in specific liquids. If, however, it is exposed to ultraviolet radiation (u.v.) it polymerizes (i.e. cross-linkage bonds between its molecules are formed) and it becomes insoluble. Therefore, if the photoresist is exposed to u.v. through a working plate, the use of the appropriate solvent (the specific liquid) removes the photoresist film wherever the working plate did not permit the passage of the u.v. The wafer, with its patterned photoresist film, hardened by subsequent heating, is then placed in a solution of hydrofluoric acid. This acid will dissolve this silicon dioxide film wherever it is not protected by the layer of photoresist on it. This acid does not, however, attack either the photoresist or the silicon. Finally, after rinsing and drying, the photoresist is removed by a different chemical solvent.

The acceptor boron or the donor phosphorus are then diffused from the vapour state into the silicon exposed by the patterned silicon dioxide film. All this integrated circuit manufacture has to be undertaken in an ultra-clean room — suffice it to state that the number of dust particles (likely to abrade the i.c.) per unit volume of the air is arranged to be not more than 1 per cent of that in a nominally clean hospital laboratory.

An alternative method of rapidly growing importance to the last stages of this process of i.c. fabrication is to utilize *ion implantation*.

The dopant atoms (and arsenic can now be included, if required) are ionized (usually positively) and accelerated through a p.d. of several hundreds of kilovolts. The high-energy beam of ions formed is directed in a vacuum with a measured beam current and for a known time (the product of this beam current and time decides the amount of the dopant implanted) on to the silicon wafer where ions become buried to various depths depending on the nature of the element and the energy of the ions. Subsequently, the silicon crystal has to be tempered by heating to remove the damage to the crystal lattice brought about by the energetic ion implantation. Whereas a diffusion process may take up to 24 hours, the doping element can be introduced as energetic ions in a tenth of that time.

When a wafer of silicon has been through the first stages of manufacture indicated above by the use of either chemical vapour diffusion or ion implantation (sometimes both are used), electrically conducting leads have to be produced between the components and the lands. These conducting leads and lands are the uppermost film on the hundreds of i.c. chips in the silicon wafer. They are usually made on the wafer by deposition consequent upon the evaporation of aluminium in a vacuum. The heating of the aluminium in a crucible so that it melts and vaporizes readily in the vacuum is done by an electron-beam gun source. The deposition on the wafer surface of this aluminium is through a second set of appropriately patterned working plates. The final wafers are subjected to a series of well-designed, computer-controlled, electrical tests. The wafer is then fashioned into the chips by breaking along the previously diamond-scribed lines, where each chip contains the whole of an integrated circuit (Figure 6.2(a) is a much enlarged photograph of an integrated circuit at this stage). It remains to insert this integrated circuit in a suitable housing of an insulating material provided with terminal pins. Furthermore, the conducting lands (to which the leads are connected) have to be joined by specialized micro-welding (cold welding) methods to wires (usually of gold of 0.001 inch diameter) which are brought out to the terminal pins. As the housing (encapsulation) is usually a black plastic, the Americans — with their fondness of the apt word — call the resulting housed integrated circuit a 'bug' (see Figure 6.2(b)). However, those used in space vehicles and in communications satellites are mounted in a ceramic.

In integrated circuits, the dimensions of the components and the separations between them are becoming so small (especially in VLSI) that the limit of resolution of the optical lenses used in their production is beginning to become critical. To overcome this problem, the technology which is developing rapidly and is most promising is that of electron-beam lithography (EBL). In this process, a fine beam of energetic electrons is made to scan the silicon wafer and

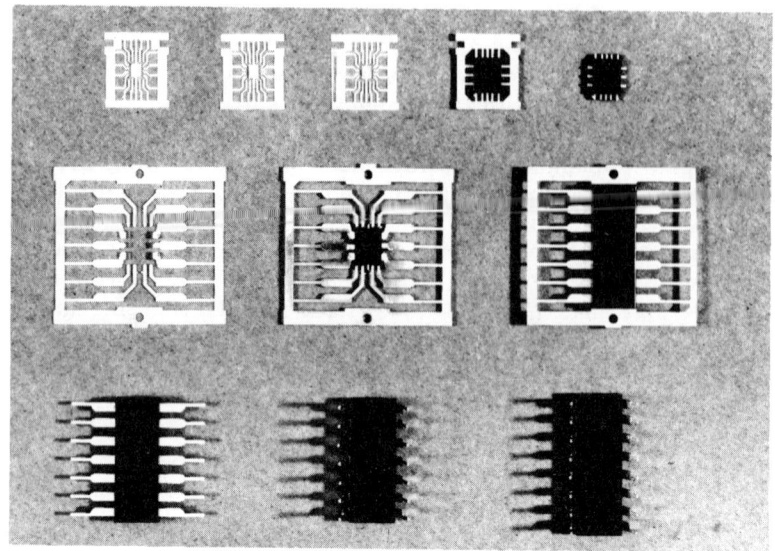

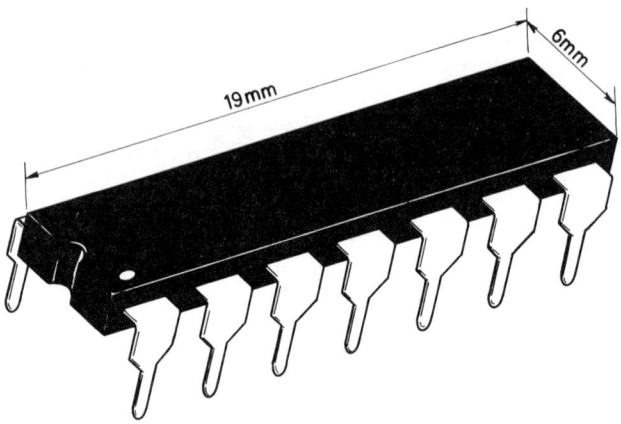

Figure 6.2. (a) Photograph of successive stages in the encapsulation of an integrated circuit dual-in-line package, and (b) a 'bug', i.e. an encapsulated integrated circuit which has 14 pins. Some have as many as 40 pins.

an electronic resist is used instead of a photoresist. The main advantage of EBL — apart from the ability to crowd even more components (usually MOSFETs as the active components) in a given area — is that the required pattern of the integrated circuit can be written directly into each chip in the silicon wafer from the information stored in the digital computer; masking is not necessary. This is because this digital information is readily converted into the required voltage variations needed to alter the position of the focus of the electron beam and the beam intensity. The main disadvantage is that the silicon wafer has to be scanned by the electrons instead of being exposed all at once (a problem which is being solved by recent developments) as in the use of u.v. The use of EBL promises to make available integrated circuit chips each containing a total of several 100 000 components.

Exercise 6

1. Give statements which define concisely the terms: (*a*) active component; (*b*) passive component. Also name three examples of each type of component.

2. A MOSFET makes use of a silicon dioxide film of thickness 100 nm (1 nm $= 10^{-9}$ m) between its gate electrode and the underlying substrate. Given that the relative permittivity ε_r (i.e. the dielectric constant) of silicon dioxide is 4.5, calculate the capacitance C which would be present due to this film between a gate of area $S = 1$ mm^2 and the underlying substrate. The appropriate equation for this capacitance C in farad is

$$C = \varepsilon_r \varepsilon_0 \ S/d$$

where ε_0 is the permittivity of free space $= 8.854$ pF/m, and d is the separation in metre between the gate electrode and the substrate (taken to be the thickness of the silicon dioxide film).

  Given that the time constant T for a capacitance C in conjunction with a resistance R is $T = CR$ where T is in second with C in farad and R in ohm, calculate the time constant involved with such a MOSFET in an integrated circuit where the area S is 10 square micrometre and R is 1 MΩ ($= 10^6$ ohm).

3. Describe in outline the chief stages in the fabrication of an integrated circuit which is based on n-p-n bipolar transistors.

4. Write short notes on each of the following:
   (*a*) Silicon is the most important semiconducting element particularly in the fabrication of integrated circuits.
   (*b*) SSI, MSI, LSI and VLSI.
   (*c*) The simple principles of two ways in which, where required, electrical insulation is provided in an integrated circuit.
   (*d*) The provision in integrated circuits of passive components.

5. What is meant by *ion implantation*? Give two advantages that it affords over the use of chemical vapour diffusion in the manufacture of integrated circuits.

6. Write an account of the principles involved in electron-beam lithography. Give one advantage and one disadvantage of this method of making integrated circuits.

7 Amplifiers: especially operational amplifiers and some of their applications

7.1 Introduction and some Recapitulation

An amplifier is a piece of equipment which has an input and an output and the output is a magnified replica of the input. Our concern is with electronic amplifiers: in the ideal case the waveform of the output from an electronic amplifier should be the same as that of the input but be (usually) of greater magnitude. The amplifier is the most important electronic system. It is based on active components joined to passive components and to the power supply. Single-stage amplifiers based on a transistor have already been described: sections 4.6 and 4.7 deal with the use of a bipolar junction transistor as the active component in common-emitter connection; section 4.18 describes the use of a field effect transistor as the active component in a common-source amplifier. The more recent and frequently adopted practice (which it is difficult — if not impossible — to improve upon) is to use an integrated circuit amplifier (except for power amplification (section 7.3)) in which the necessary active and passive components are within an i.c. of the monolithic type. The widely used operational amplifier is a very important case of an integrated circuit.

The term 'signal' is used to denote either a current or a voltage (usually the latter) which may be steady or varying with time. The input signal to an amplifier is usually small; it controls the output. In the general case, the input signal is denoted by the symbol X_i and the output by X_o. The most frequently encountered case is where the input signal is a voltage variation; in which case it will be denoted by v_i and the output is also a voltage variation denoted by v_o.

The *gain of an amplifier* is the ratio of the magnitude of the output to that of the input signal. In general, therefore

$$A = X_o/X_i \tag{7.1}$$

In the case of a voltage amplifier

$$A_v = v_o/v_i \tag{7.2}$$

where A_v is the voltage gain. For a current amplifier

$$A_i = i_o/i_i \qquad (7.3)$$

where A_i is the current gain.

7.2 Amplifiers: Classified according to the Nature of the Input Signal

The input signal voltage may be:

(a) a change of voltage from one steady value to another, involving a change, ΔV_i. The purpose of the amplifier is then to provide an output voltage V_o which is changed by ΔV_o which is much larger than ΔV_i. Indeed, equation (7.2) becomes

$$A_v = \Delta V_o/\Delta V_i \qquad (7.4)$$

(b) The input signal voltage is alternating and is represented by a sine wave (it is said to be 'sinusoidal') of a single frequency $f = \omega/2\pi$ and

$$v_i = V_p \sin \omega t$$

where V_p is the peak value of v_i.

This is the case of an alternating voltage amplifier. The implication here is that the input signal is alternating at a single frequency f. By the use of Fourier analysis, it can be shown that all periodically varying voltages (whatever their waveform shapes) can be represented by the summation of a series of sine and cosine waveforms of frequencies which are simple integral multiples of a fundamental frequency and are of various peak values.

(c) The input signal v_i is a *pulse*. Here the concern is the very important *pulse amplifier*. The input is present for a very short time and then may be repeated after a lapse of time. The output is to be a larger pulse of the same shape (as far as possible) as that of the input pulse. The 'rise time', t_r, 'fall-time' t_f and 'width' of the pulse (Figure 7.1) are important factors in arriving at a suitable design of amplifier.

7.3 Power Amplifiers

In this case the output signal, X_O, is a fairly large power output, P_O, because this output has to drive some kind of electromechanical device which converts the output electrical information from the amplifier into corresponding mechanical forces. A common example of a low resistance electromechanical device is a moving-coil loudspeaker. As the resistance is low, large current changes of the output are needed to ensure that P_O is big enough to provide enough

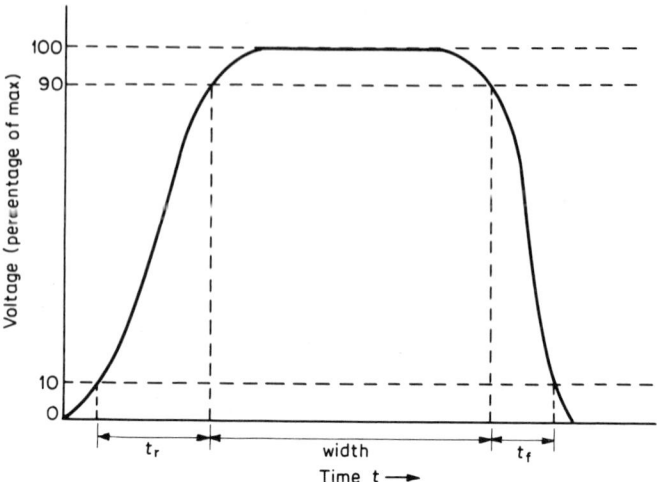

Figure 7.1. A voltage pulse

sound in the room, where

$$P_O = I_{Orms}^2 R$$

R being the resistance of a moving-coil loudspeaker and I_{Orms}, the root-mean-square (r.m.s.) value of the output current.

The input to the amplifier is a voltage signal of small amplitude (usually) from such a device as a microphone or a gramophone pick-up or a magnetic-tape cassette player. This input is first to a pre-amplifier stage, then to a second stage whereby the varying voltage (from the output of the pre-amplifier stage with the microphone, say, as the input) is amplified to be large enough to act as the varying voltage input (the larger signal needed) to the power amplifier stage, the output of which is feeding the loudspeaker as its load.

The audio-frequency (AF) amplifier involved between the small voltage input and the final power output is thus usually a three-stage amplifier. A moving-coil loudspeaker as the electrical load in the power output stage of such an audio-frequency amplifier may have a resistance of only 8 ohm and yet need to be operated by an electrical power of 20 watt. So the output stage must be able to provide considerable power requiring fairly large alternating currents. Also, the load resistance of the electromechanical device must be matched to the output stage for optimum performance, i.e. the load on the output stage has to be of the correct resistance (see section 4.6).

There is clearly a demand for transistors able to provide a power output of 20 watts or more. These are supplied as individual components, (not as part of an i.c.) with 'heat sinks' attached to ensure

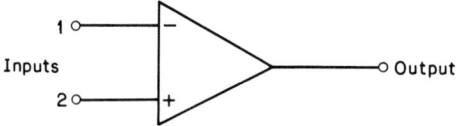

Figure 7.2. Circuit symbol for a difference amplifier. *Note*: the − sign indicates that 1 is the inverting terminal; the + sign indicates that 2 is the non-inverting one

against any excessive rise of temperature. This demand has been met by discrete bipolar junction transistors and by the recent n-channel enhancement mode power MOSFET.

7.4 The Prime Function of an Amplifier

In all amplifiers, the output power is derived from the power supply, which is a battery or a rectifier unit operating from the mains supply. The amplifier serves as a control device in that the input signal to it controls the power so that its waveform is a replica of that of the input signal. Essentially, it is power that is concerned. Whether voltage or current amplification predominates is dependent upon the values of the input and the output passive components and the characteristics of the transistor used as the active component.

7.5 Difference or Differential Amplifiers

This very useful type of amplifier has two input signals (usually voltages: V_1 and V_2). The symbol for such a difference amplifier (Figure 7.2) is an obvious development from the triangular symbol (Figure 4.20) used for an amplifier in general. The control of the output voltage, V_O, is achieved by the difference between the input signal voltages to the terminals 1 and 2, i.e. by $(V_1 - V_2)$.* The design is such that the output is affected as little as possible by the mean input voltage $(V_1 + V_2)/2$. If at all possible, a differential amplifier — often an operational amplifier — should be chosen for a given task in electronics.

As the difference amplifier (known as the DIFF. AMP.) has its output decided almost entirely by $(V_1 - V_2)$ and not by $(V_1 + V_2)/2$, this ability to ignore (ideally, completely; in practice, almost) the so-called *common-mode signal* $[(V_1 + V_2)/2]$ is of great value in the amplification of input voltages. This is because there is frequently present at inputs a voltage variation which is not wanted — the amplifier should reject it. Indeed, the ability of a difference amplifier to ignore the common-mode signal is decided by its common-mode rejection ratio (CMRR)

---

* Note that $(V_1 - V_2)$ is involved and not $(V_2 - V_1)$ because the terminal 1 (to which V_1 is applied) is the inverting terminal: it produces an output which goes positive as the input goes negative, and *vice versa*.

$= \rho$, where ρ is defined by the simple equation

$$\rho = A_D/A_C \qquad (7.5)$$

where A_D is the gain of the difference amplifier for the difference input signal $(V_1 - V_2)$ and is called the *differential gain*, whereas A_C is the gain of this amplifier for the common-mode signal $[(V_1 + V_2)/2]$ and is called the common-mode gain. Thus, in words

> *the common-mode rejection ratio is the ratio of the differential gain to the common-mode gain of the amplifier.*

For the difference signal, the output is $A_D(V_1 - V_2)$. For the common-mode signal, the output is $A_C(V_1 + V_2)/2$. In order for the first of these two outputs to be much greater than the second

$$2A_D(V_1 - V_2)/A_C(V_1 + V_2)$$

must be large.

This means that A_D/A_C must be large, i.e. ρ must be very big. In practice, values of ρ of 10^5 are possible. This large value of ρ (i.e. CMRR) is very important when a small input signal is present on a large steady voltage level and it is necessary to amplify greatly the small input signal and have an amplifier circuit which ignores — as far as possible — the steady voltage level. Thus, it is possible to have a voltage level of 1 or 2 volt on which there is superimposed a variation, the true signal (the one which it is required to amplify) of tens of millivolts.

7.6 Operational Amplifiers

Perusal of a recent catalogue establishes that the electronic devices supplied by the various manufacturers are frequently in the form of units which are integrated circuits. Such i.c.s are usually of the monolithic type based on a silicon chip, if the power output is modest. The operational amplifiers (opamps) form an important class of such integrated circuits. Their importance is emphasized by the fact that there are about 2000 types of i.c. available commercially, of which perhaps 200 are opamps.

The opamps are multi-stage. Unlike the multi-stage amplifiers described in section 4.9 which makes use of resistance-capacitance coupling, the stages are always directly coupled together. This means that capacitances (which will only pass alternating currents) are not used, only resistors. Thus, the opamp will not only operate with a varying input signal voltage, but also with a steady (or d.c.) input signal voltage. When used as amplifiers, opamps are always provided with external components to ensure a considerable amount of negative feedback (section 4.10).

In virtually any electronic system constructed in a school or college laboratory it is impossible to achieve a performance comparable with that obtained by using an operational amplifier. The characteristics of even a low-cost opamp, such as the 741, approach those of the ideal voltage amplifier in that the 741 has:

(a) a very big open-loop gain — approximately 10^5;
(b) a very big input impedance — approximately 2 MΩ;
(c) a low output impedance — some tens of ohms.

In the 'ideal' (unattainable) case, both (a) and (b) would be infinity and (c) would be zero.

In all cases in which an operational amplifier is used as a linear amplifier (i.e. one where the output voltage is directly proportional at all times to the input voltage), negative feedback is used. This feedback (apart from reducing the voltage gain, which does not matter, see section 4.10) provides the following enormous advantages:

(i)   The linearity is excellent and the greater the degree of feedback (i.e. the larger is β in equation 4.5) the better is the linearity.

(ii) The bandwidth (defined as the range of frequencies of the input signal over which the voltage gain is nearly constant) increases with the negative feedback applied. It is hence easy to obtain a linear amplifier which operates from zero frequency (or d.c. voltage) up to 50 kHz.*

(iii) The amplifier is current limited (usually to 10 mA) so that it is not damaged if the output terminals are short-circuited.

(iv) The output resistance decreases as the negative feedback is increased. In practice, the output resistance can therefore readily be only 10 milliohm. Further advantages of opamps are given in later sections, where appropriate.

A recommended general purpose opamp for students' experiments and projects is the 741. This is supplied in an 8-pin dual-in-line (dil) plastic package, as shown in Figure 7.3(a). It is based on an input circuit comprising matched bipolar transistors in a difference amplifier. As is to be expected, a power supply is necessary. This is a stabilized supply of + 15, 0, − 15 V which is connected to specific terminals of the amplifier, as shown in Figure 7.3(a). The 741 opamp will operated satisfactorily with power supply voltages which are as low as perhaps + 7, 0, − 7 V. The two voltages (nominally + 15 V and − 15 V) are with respect to the common line (usually, but not necessarily, earthed) at 0 V. The use of two supply voltages is essential with a difference amplifier to enable amplification of an input signal

---

* Many opamps (including the 741) have internal frequency compensation to ensure this.

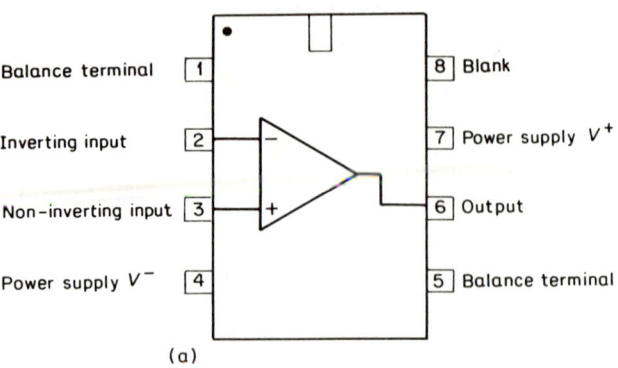

(a)

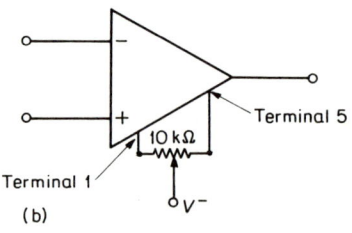

(b)

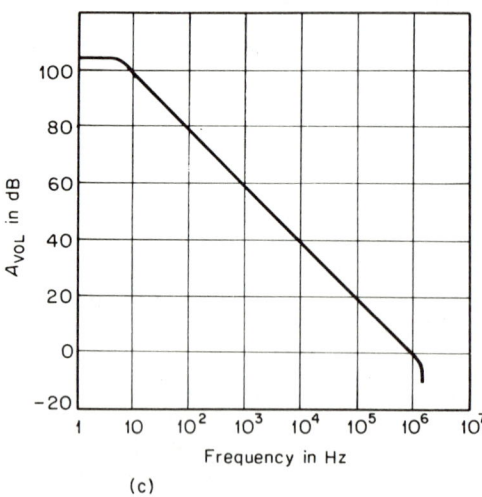

(c)

Figure 7.3. The 741 operational amplifier (a) the dil plastic package (top view), (b) the connection of the variable resistor R for zeroing, and (c) graph of open-loop voltage gain (A_{VOL}) vs. frequency in Hz A_{VOL} in dB

which may be steady and either positive or negative with respect to the common line. The usual practice is to omit this power supply from the circuit diagram in which the opamp appears. But it *must* be there. The commonest fault of the beginner in the use of opamps in practice is to forget to connect the zero terminal of the power supply to the common line of the amplifier.

Operational amplifiers are almost always based on a difference amplifier input stage comprising two transistors. These transistors may be of the field effect type instead of bipolars as in the 741 (which has an input resistance of 2 MΩ). For many applications in which a particularly high input resistance is required a MOSFET-based input DIFF AMP is used in the i.c. Available models of this MOSFET input type are the 31305S and the 3140S. There is also a useful E78 opamp which has an input resistance about 5000 times that of the 741: the E78 makes use of a dual JGFET input circuit; it is not an i.c., but a discrete modular unit in which the individual circuit components are mounted on a small printed circuit board and then encapsulated in a plastic housing to provide a unit of about $25 \times 25 \times 13$ mm.

As the input to a 741 is a DIFF AMP and because it is virtually impossible to have two transistors in the input circuit which are an exactly matched pair, so the input voltages set at $V_1 = V_2 = 0$ do *not* produce a zero output voltage. This is undesirable. This problem is easily solved by the use of an external compensating network for this input off-set voltage. This may well involve only one external resistor R (Figure 7.3(b)).

The output current which the 741 opamp can supply is limited to 10 mA, so that the amplifier is not damaged if the amplifier output terminals are short-circuited.

Figure (7.3(c)) is a plot for a 741 of the voltage gain (in dB) against the frequency of the input alternating voltage. The voltage gain is the so-called *open-loop voltage gain* (A_{VOL}), which is the gain provided when the feedback is absent. To obtain this response, the 741 opamp is provided internally with frequency compensation (as stated in the footnote to p. 157).

For the 741, some characteristics are given on p. 157. Ideally, the bias input current and off-set voltage should both be zero. For the 741, these values are, respectively, 20 nA approximately and 1 mV approximately.

Operational amplifiers almost always make use of negative feedback (section 4.10). This is provided by means of two resistors, R_1 and R_2. Two ways of connecting these are used, as shown in Figure 7.4. The − sign is the input to the inverting amplifier whereas the + sign indicates the input to the non-inverting amplifier. Figure 7.4(a) is with the input voltage applied to the − terminal (via the resistance R_1) and the common-line and Figure 7.4(b) is with the input applied

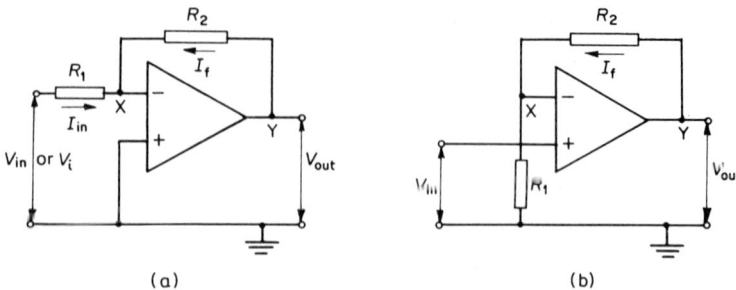

Figure 7.4. Two ways of providing negative feedback to an opamp: (a) the inverting configuration, and (b) the non-inverting configuration

between the + terminal and the common-line. Circuit (a) is in the so-called *inverting configuration*; circuit (b) is in the *non-inverting* configuration. As shown below, the p.d. between the point X and the common-line is virtually zero at all times. Thus, in the circuit of Figure (7.4(a)) the input resistance to the amplifier is R_1 whereas that of the circuit (b) is ideally infinite (for the 741 it is 2 MΩ).

When the feedback is *not* applied, i.e. the resistors R_1 and R_2 are absent, the voltage gain of the amplifier is said to be open-loop and is designated by A_{VOL}. When the feedback *is* applied, i.e. the resistors R_1 and R_2 are present, the voltage gain is said to be *closed-loop* and is designated by A_{VCL}. These terms and symbols are used in all cases of amplifiers, without or with feedback; they are by no means reserved only for use in describing opamps.

The opamp connected as in Figure 7.4(a) is such that the output voltage V_o is in antiphase with the input voltage (cf. section 4.5). This is why the configuration is called 'inverting' and why the upper input terminal in the triangle is marked with a negative sign to denote that it is the inverting terminal. When the potential at point X tends to go positive, that at the output terminal Y goes negative, and vice versa. A_{VOL} is big (it is 10^5 for the 741 and this is not a large number for an opamp). So any possible change with respect to the common-line of the potential at X due to a change of the input current I_i (brought about by a change of V_i) would be accompanied by a large change (in the opposite direction) of the potential at the point Y with respect to the common-line (earth, usually). The result is that the feedback current I_f through the resistor R_2 will be equal at all times to the input current I_i, whether V_i increases positively or negatively. This means that the p.d. V_X between X and the common-line is maintained at zero (within a few microvolts or so) by the negative feedback. Indeed

$$V_X = 0$$

and X is called a 'virtual earth' point because there is virtually no voltage difference between the two inputs.

In the non-inverting configuration, V_X will not be zero as the same considerations do not apply. In Figure 7.4(a), $V_X = 0$ because the non-inverting input is connected to earth. In Figure 7.4(b), $V_X = V_{in}$. In the first circuit (Figure 7.4(a)) because no current flows into the amplifier

$$I_i = V_i/R_1 = I_f = -V_o/R_2$$

$$A_{VCL} = V_o/V_i = -R_2/R_1$$

(7.6)

In the second circuit (Figure 7.4(b))

$$V_i = V_X = V_o R_1/(R_1 + R_2)$$

$$A_{VCL} = V_o/V_i = (R_1 + R_2)/R_1 = 1 + R_2/R_1$$

(7.7)

The possible high open-loop voltage gain, A_{VOL}, of the operational amplifier means that the negative feedback provided can be large so that excellent linearity of the gain is achieved over a range of values of the frequency. Also, a high input resistance is obtained which can be very high indeed for a good quality opamp in the circuit of Figure 7.4(b).

7.7 Typical Characteristics of an Operational Amplifier in the Inverting Configuration

In the inverting configuration (Figure 7.4(a)) the graph of the output voltage, V_o, against the input voltage, V_i, is linear up to values of V_o of approximately ± 14 V, when saturation sets in (Figure 7.5(a)). By linear is meant here a straight line; by saturation is meant no further increase of the dependent variable (in this case, V_o) with increase of the independent variable (V_i). This saturation is to be expected when the power supply (as usual) is $+15$V, 0, -15V.

In Figure 7.5(b) is shown a plot of the closed-loop voltage gain, A_{VCL}, against the frequency of the input signal with $R_2/R_1 = 100$, and the circuit of Figure 7.4(a) is used. The gain drops at a frequency of 10^4 Hz. The bandwidth increases with increased negative feedback and hence lower gain.

7.8 Some Applications of Operational Amplifiers

There are many applications of opamps. Here and in the following section (7.9) only a few are given.

Operational amplifier as a voltage follower

The circuit of Figure 7.6(a) produces an output of exactly the same waveform and magnitude as the input signal, i.e. $V_o = V_i$. The role of

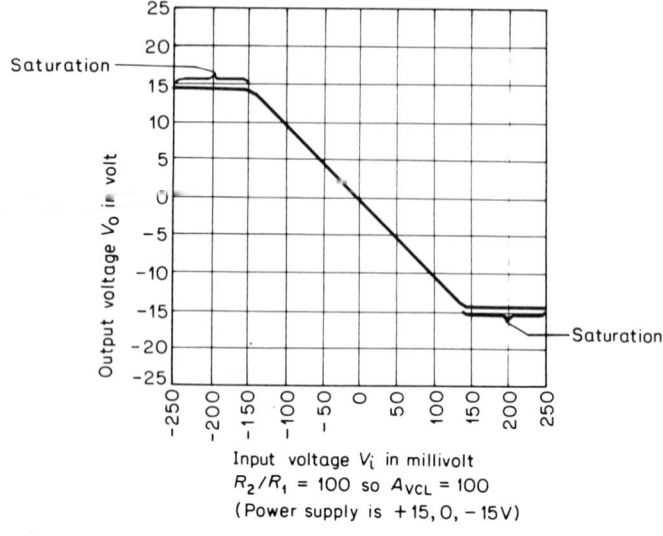

Input voltage V_i in millivolt
$R_2/R_1 = 100$ so $A_{VCL} = 100$
(Power supply is $+15, 0, -15V$)

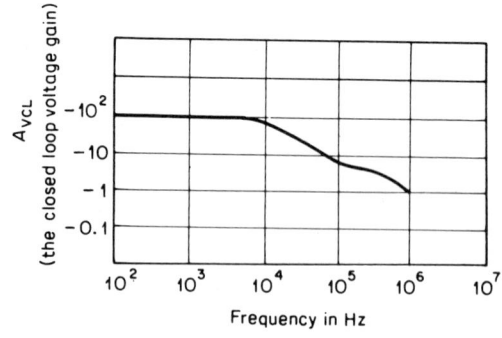

Frequency in Hz

$R_2/R_1 = 100$ so $A_{VCL} = -100$

Figure 7.5. Typical characteristics of an operational amplifier in the inverting configuration: (a) measured output voltage, V_o, against measured input voltage, V_i, (b) closed loop gain A_{VCL} against input signal frequency

this unity gain amplifier is important as it functions as an impedance changing device: it is sometimes called a buffer amplifier. The input resistance to the non-inverting stage of an operational amplifier is very high. Suppose it is $10^9 \, \Omega = 1000$ MΩ. R is chosen to be, say, 10 MΩ; this is so much smaller than 1000 MΩ that it decides the input resistance.

The output resistance of the amplifier stage is tens of milliohms, when negative feedback is applied. Hence, the output signal (which is identical with the input signal) is now available from a low resistance

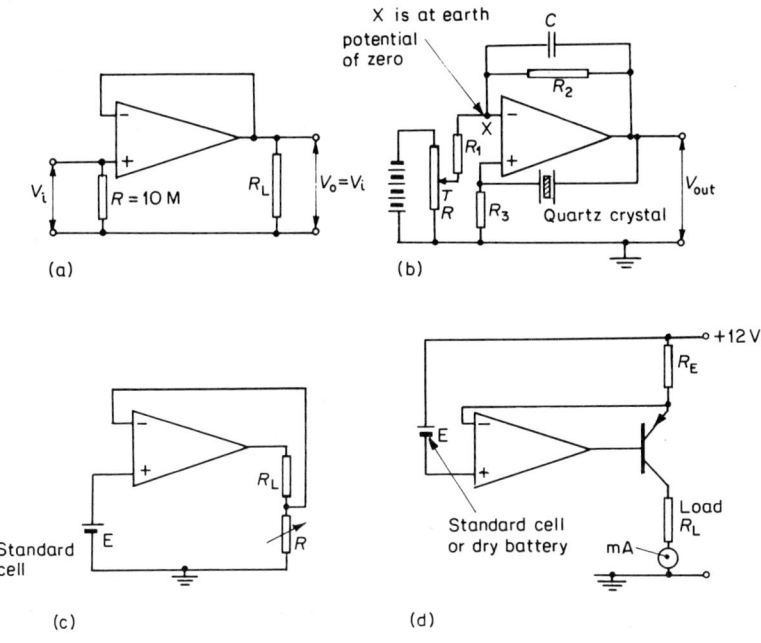

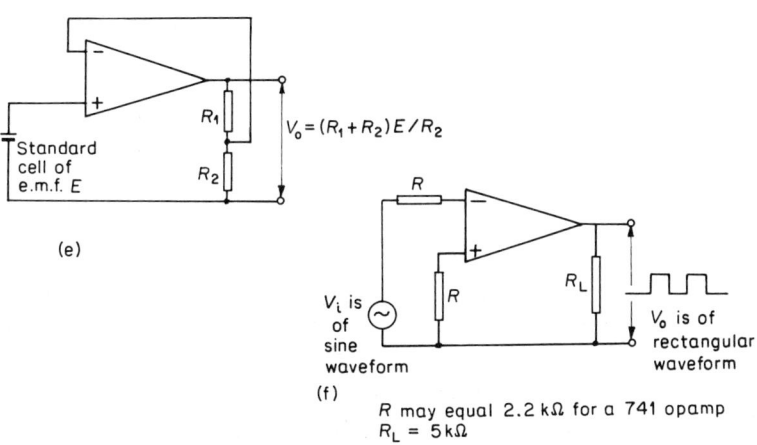

Figure 7.6. Some applications of operational amplifiers (a) as a voltage follower, (b) a crystal-controlled oscillator based on an operational amplifier, (c) in a constant current source, (d) an alternative more complex but improved circuit to that of (c), (e) in a constant voltage source circuit, and (f) to obtain a rectangular waveform from a sinusoidal voltage waveform

amplifier stage although it originated from a source of very high resistance. This is a valuable circuit arrangement. Conventional amplifiers can now be used to amplify this signal.

This voltage-follower network may be regarded as a current amplifier because the current through the load resistance across the output is many times the input current yet there is no gain of voltage. The output current is, in fact, determined by the load resistance connected across the output terminals (assuming that the current drain in the opamp does not exceed 10 mA).

A crystal-controlled oscillator based on an operational amplifier

In section 4.10, dealing with feedback, it is stated that positive feedback is employed in oscillators. In section 4.14 are described three types of sinusoidal oscillator (i.e. circuits which provide an output which is continuous and of sine waveform) which are based on the amplification provided by a bipolar transistor and the use of positive feedback. These three could all be based on the use of an operational amplifier. Only the most important of the three is considered here: it is the crystal-controlled oscillator given in Figure 4.25 in which the active component is an n-p-n bipolar transistor. In Figure 7.6(b) is shown the circuit diagram of a crystal-controlled oscillator based on an operational amplifier. Note that the necessary positive feedback is provided via the quartz crystal (as before, a 465 kHz one) and is developed across the resistor R_3 which is connected to the non-inverting input terminal of the opamp. R_3 should have a value approximately equal to $R_1 R_2/(R_1 + R_2)$. To prevent the occurrence of the negative feedback of alternating current, the capacitor C is joined across the resistor R_2. The reactance of C is $1/2\pi f C$, where the frequency f is, say, 465 kHz. This reactance should be less than $R_2/500$. Note that the frequency f of the output provided by a quartz crystal-controlled oscillator is fixed. It is also extremely constant.

It is, of course, possible to build circuits based on an operational amplifier which correspond to the other two oscillators described in section 4.14. This can — if desired — be undertaken as a project. No further details are given here because it is now preferable to purchase integrated circuit oscillators and, indeed, integrated circuit voltage stabilizers rather than construct them oneself by making use of opamps and passive components.

Use of an operational amplifier as a constant current source

The simplest circuit is one where the opamp is used in the non-inverting mode and the input to the non-inverting terminal ($+$) is from a standard cell, a dry cell or a Mallory cell (Figure 7.6(c)) or a Zener diode stabilized supply of voltage (section 3.16). Even a simple dry cell can be used to produce a very effective constant current source.

The input resistance of the opamp in the non-inverting mode is so high that the current drain on the cell used at this input is negligibly small. As the negative input terminal $(-)$ is the virtual earth, the voltage across the variable resistor R in Figure 7.6(c) is equal to the e.m.f. of the cell. The total loading on the operational amplifier (if it is a 741) needs to exceed 1500 ohm to ensure satisfactory working, i.e. $R_L + R > 1500$ ohm. The current through the resistor R is E/R, where E is the e.m.f. of the cell used. There are two disadvantages of this constant current source:

(i) the constant current is limited to 10 mA;
(ii) one end of the load resistor R_L is not conveniently at earth potential.

An alternative opamp circuit which is a bit more complicated in that it includes also a p-n-p bipolar transistor is shown in Figure 7.6(d). This circuit makes use of 100 per cent negative feedback. The reference cell connected to the opamp non-inverting input will now drive the p-n-p bipolar transistor into a conducting state. The current through the resistor R_E will increase until the p.d. across it is equal exactly to the e.m.f. E of the reference cell. In the setting up of this improved circuit, the p-n-p transistor has to be chosen so that it is able to carry current in excess of the constant current required; it is mounted on a heat-sink if it is likely to overheat. Two resistors can be used for R_E: one is fixed at, say, 100 Ω to limit the current through the bipolar transistor and the second one is in series and variable to select the constant current needed.

Example 7.8

Based on a Mallory cell of e.m.f. 1.35 V as a reference voltage, the circuit of a constant current source (as in Figure 7.6(d)) to provide 50 mA is needed. What is the approximate value of the required resistor R_E? Write a brief note on the provision of R_E, in practice.

$$1.35 = R_E \times 5 \times 10^{-2}$$

$$R_E = 27 \ \Omega$$

R_E could be a fixed 20 Ω resistor in series with a variable 10 Ω resistor for accurate setting of the required amount. Both these resistors comprising R_E should preferably be wire-wound so that they can carry currents in excess of that required without overheating. Any resistance change of R_E with temperature would alter the current through the load.

Use of an operational amplifier as a constant voltage source

An appropriate circuit is shown in Figure 7.6(e). This is much the same as that of Figure 7.6(c). The current provided by the opamp will arrange to keep the p.d. between its input terminals $(-$ and $+)$ equal to zero. Thus, the voltage input $V_i (= E)$ appears across the resistor R_2.

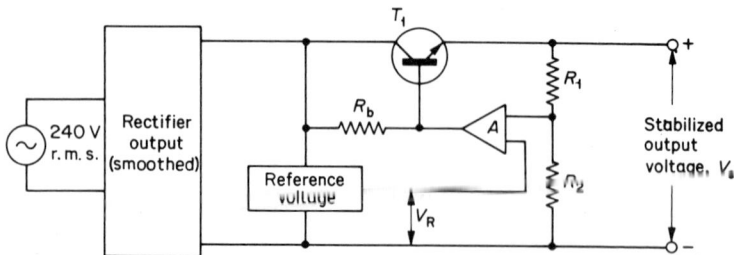

Figure 7.7. A stabilized voltage supply utilizing an operational amplifier. (The transistor T_1, or a Darlington pair, form the series regulating element. All the load current passes through this regulator. As the intention is to maintain constant the output voltage V_o, there must be maintained across the series regulator the difference in voltage between the input (the unstabilized supply) and the output voltage.

The output voltage V_o —which can be larger than the cell voltage E — appears across R_1 and R_2 in series and is given by

$$V_o = E(R_1 + R_2)/R_2$$

Use of an opamp to obtain a rectangular waveform from a sinusoidal voltage waveform

A feedback resistor R_2 is not used, so that the opamp is working in 'open-loop' (Figure 7.6(f)). The voltage input (to the inverting terminal) has a sine waveform with positive and negative values sufficient to drive the opamp from negative to positive saturation at a frequency decided by this input signal (cf. the characteristic given in Figure 7.5(a)).

7.9 A Practical Power Unit with a Highly Stabilized Voltage Output based on an Operational Amplifier and a Series Transistor

The stabilised voltage supply unit shown in Figure 7.7 requires the use of an n-p-n bipolar transistor for series regulation. As regards voltage stabilization, the performance of such a circuit is improved considerably if an operational amplifier (essentially a difference amplifier here) is used as shown in Figure 7.7. In the Appendix there is given the circuit diagram of a useful multimeter, based on an opamp.

Exercise 7

1. In connection with electronic amplifiers, give definitions of the following terms: *single stage*; *multi-stage*; *gain* (including voltage gain, current gain and power gain).
2. Define four classes of amplifier according to the nature of their input signals and outputs. What is an audio–frequency amplifier? Explain the

idea and the importance of Fourier analysis of waveforms in connection with amplification.

3. What is a difference amplifier? Why is such an amplifier an important device in electronics? In connection with such amplifiers, define the following terms: (a) the common-mode signal; (b) the common-mode gain; (c) the common-mode rejection ratio.

4. In the case of operational amplifiers, write accounts of:
   (a) the nature of the supply voltage used (circuit diagrams are not needed);
   (b) the feedback circuits.

5. Derive for an operational amplifier, the equations:
   (a) $A_{VCL} = -R_2/R_1$ in the inverting configuration,
   (b) $A_{VCL} = 1 + R_2/R_1$ in the non-inverting configuration.
   A_{VCL} is the closed-loop voltage gain, R_1 and R_2 are the resistors connected as in the basic circuit diagram of Figure 7.4.

6. Draw a circuit diagram and explain the working of a stabilized voltage supply which makes use of a difference amplifier in the form of an operational amplifier.

7. Draw the circuit diagram and explain the working of a quartz crystal-controlled oscillator which produces a sine wave output of frequency 465 kHz and makes use of an operational amplifier.

8 The fundamentals of digital circuits and some applications

8.1 Analogue and Digital Methods

An *analogue* method is one in which the input to some device is continuous (it is present all the time the apparatus is switched on) and the output of the device is likewise continuous. The familiar example in everyday life is the watch or clock with hands which rotate around the centre of a circular face marked in hours or minutes. Another familiar device in the home is the record player. The record itself may be a disc (with a pick-up) or a magnetic tape in a cassette (with an arrangement for converting the magnetic variations on the tape — which represent the sound — to a varying e.m.f.). The input to an amplifier in the record player is on continuously. This amplifier is an audio-amplifier in that it responds satisfactorily to input signals over the audio-frequency range. The output from this amplifier is continuously present and varies in the same way as the input signal; it operates the loudspeaker which emits the sound.

There are analogue computers. They have input data which vary continuously and a corresponding output. Indeed, the operational amplifiers (Chapter 7) are much used in analogue computers. The analogue computer, in one of its most advanced forms, was developed to simulate the flight behaviour of an aircraft, so is valuable for air-pilot training.

The old moving-coil ammeters and voltmeters with their 'pointers' which move around a calibrated dial are analogue instruments: the input is the direct current or alternating current (depending in the latter case on the presence of a rectifier in series with the moving-coil meter; note that a meter for recording voltage is in both cases a current meter supplied with a multiplier) and is present continuously; the output is also present continuously, it is the position of the 'pointer' on the calibrated dial. The slide-rule is an analogue device.

The revolution which has taken place in electronics during the past 20 years or more has been the enormous development of digital techniques as compared with analogue ones. The developments (which continue still) in integrated circuits (Chapter 6) have resulted in microelectronics; this has given us digital circuits which can operate at very high speeds, in limited space and relatively cheaply. Pulses which last only a few nanoseconds (1 nanosecond = 1 ns = 10^{-9} s) separated by similar times are used in fast digital electronics. This means that on/off information (i.e. binary data as used in computing; see sections 8.2 and 8.17 can be supplied so fast that 50 on/offs (on and off are both in the form of the presence or absence of an electrical voltage which may be present for only 10 ns) can be conveyed in only one microsecond, μs (1 μs = 10^{-6} s). This revolution has provided electronic calculators, the digital computer, the minicomputer and the microprocessor. The analogue computer has now been largely replaced by digital devices, though some 'hybrid computers' are still used which incorporate both analogue and digital methods.

Accompanying this enormous progress, which has resulted in the digital computer, there have been developed devices such as digital watches and clocks (especially the battery-operated quartz clock, so-called) which record the time in digits as 09.58, or whatever. The familiar moving-coil electrical measuring instruments have been replaced by the digital voltmeter (DVM) and the digital multimeter. Moreover, there has been a whole new development in digital control practice in applied science and technology. The tendency here has been to use the microprocessor more and more.

Telecommunication — especially over great distances — has been greatly improved by the use of digital methods, though the domestic radio receiver is still analogue. Within telecommunication is included the conveyance of both audio and visual information (television being the chief example of the latter) and the use of communication satellites.

It must suffice here to stress only one further case of the advantages of digital over analogue methods. It is that an electronic circuit used in analogue fashion will provide an output which will vary over a period of time because it may be subject to variations of the input, of its power supplies and changes of the characteristics of its components due to their ageing or replacement. Transistors of a given type are notoriously difficult to make all alike, though in integrated circuit form the variations are not so great as between different models of the same type in discrete form. A digital circuit will have an input in the form of a presence or an absence of voltage (usually) so that the corresponding output from the digital circuit is either present or not. The device being controlled by digital electronic apparatus is thus not as subject to variations due to changes of the power supply (other than complete failure thereof) or changes in the magnitudes of electronic components due to their ageing or replacement. In the development of digital electronics, and especially of the digital computer and its by-products, a most notable part has been played by the use of logic gates (section 8.2) and the generation of pulses by circuits of the multivibrator type (section 8.15).

8.2 Logic Gates

These are two-state systems in that they are either ON or OFF. The simplest two-state system is an ON–OFF switch. This is either open or closed. Although the simple switch or such switches in simple series or simple parallel circuits can be used to conduct most of the logical operations required, they cannot go from the ON to the OFF state (or vice versa) rapidly enough for modern digital circuits. With an electronic switch or gate it is possible to do this in 10^{-8} s (10 ns).

The importance of the logic gate in modern digital electronics arises in the first place from its widespread use in the automatic digital computer. In this connection, all types of digital computer operate essentially with so-called binary arithmetic, based on 1 and 0 where 1 is simulated by the ON of the switch and 0 by the OFF. The adjective 'logic' is used because '1' or ON can also be regarded as 'true' whereas '0' or OFF is regarded as 'false'. Logic, after all, is concerned with what is 'true' and what is 'false'!

It is a salutary observation that those who speak of automatic digital computers as 'electronic brains', seem to overlook the fact that all the computer can do is decide extremely rapidly (100 million times per second) between these ON and OFF (1 and 0, respectively) states. With such additions as a 'memory store' and other devices, the important question is 'can the human brain do any better?'

Modern electronic gates are essentially electronic switches operated by input pulses of voltage which give rise to output voltage pulses, usually of about the same size. ON is usually when the voltage pulse height is above some specified level whereas OFF is when this pulse height is below some specified level. To be quite sure that the ON and OFF states of the electronic gate are definite, there is a significant difference in voltage between the lowest value of the upper level for the ON and the highest value of the lower level for the OFF.

There are three basic logic gates, which are AND, OR and NOT. There is also a pair of gates resulting from combinations of two of these which are so important in digital computer practice as to be effectively the two that really matter. They are NOT AND (or NAND) and NOT OR (or NOR). These are respectively combinations of the NOT and the AND operations and the NOT and the OR operations. To stress their importance, it can be added that *all* the electronic circuits of the vital central processor unit of a digital computer (section 8.18) may be based on NAND gates (or NOR gates) alone.

The EXCLUSIVE–OR gate should be mentioned; to distinguish this from the OR gate, the latter is sometimes called the

INCLUSIVE–OR. Finally, there is a valuable so-called ENABLE gate.

Electronic gates of all the seven types mentioned — but really there are only three fundamental ones — can be based on the bipolar junction transistor as in the famous transistor-transistor logic (TTL) series 74 (Texas Instruments Ltd) or in the newer metal oxide field effect transistors (MOSFETs). The latter are best used in the so-called complementary circuit involving very conveniently a gate with an n-channel MOSFET having a p-channel MOSFET as its load in its drain circuit. This is known, for short, as CMOS (sometimes COSMOS) and has led to the CMOS ('seemos') series, known as 4000. For TTL, the upper level voltage in the ON state is 2 to 5 volt and the lower level voltage in the OFF state is between 0 and 0.8 volt. For CMOS, the upper level is 70 per cent of the d.c. supply voltage and the lower level is 30 per cent of this d.c. voltage. In the case of CMOS, the supply voltage concerned may have a steady value anywhere in the range from 3 to 15 volt. Thus, for TTL, a voltage pulse of height less than 0.8 volt represents 0. Corresponding values are readily calculated for CMOS gates. Several texts are available which give details of the circuits of electronic logic gates. Such circuits are not to be studied here, because it is better to purchase and use electronic logic gates available in integrated circuit form. Thus, only the symbols for the basic three gates are given in Figure 8.1. Note that there are two families of symbols: the American (which are the more widely used) and the British (which are easier to draw and more specific). It must be mentioned, however, that CMOS gates consume much less power but are usually slower than TTL gates.

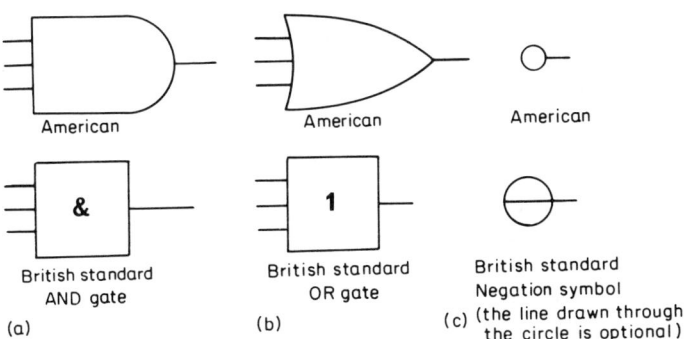

Figure 8.1. Symbols for logic gates and functions (a) AND; (b) OR; (c) the negation symbol

8.3 Binary Arithmetic and its Simulation by Logic Gates

We are used to a decimal system where the digits are 0,1,2,3,4...9. This leads, among several other features of value, to the use of the base (or radix) 10 in logarithms. The fact that the number 10 is so important in our decimal system is really only because we have 10 fingers. If any number is considered, e.g. 71 546, it is expressible in the decimal system as

$$7 \times 10^4 + 1 \times 10^3 + 5 \times 10^2 + 4 \times 10^1 + 6 \times 10^0$$

the base is 10.

If we are to use a simple logic system in electronics to represent such a number as 71 546, ON and OFF positions of a fast-acting electronic switch are available. This is all that is available in digital computers other than memory stores, input and output devices. This means that only a binary system (for which the base is 2) is possible fundamentally where the numbers are 1 and 0. The binary system had been developed and, in fact, forms the basis of Boolean algebra which was introduced by George Boole (1815–64) in his original work on logic where instead of 1 and 0, the terms 'true' and 'false' were used.

If a system of numbers can be developed based on the binary system, computation by electronics (and other) gates is possible and can be rapid. This can be done. A little thought shows that a decimal number (such as 71 546 above) can be converted into a binary one by the following procedure.

  (i) Divide the decimal number by 2. The remainder will be either 0 or 1. This is the digit on the extreme end of the right-hand side of the binary number to be obtained.

 (ii) Divide by 2 the quotient of this operation (i). The remainder gives the digit (0 or 1) in the next but last place at the right-hand end.

(iii) Divide by 2 the quotient of this operation (ii). The remainder gives the digit (0 or 1), two places from the extreme right-hand end of the binary number to be obtained.

(iv) Proceed as indicated in (i), (ii) and (iii) until the quotient obtained on division by 2 is zero, or within 1 of zero. The binary number has then been determined.

As an example, consider the decimal number first introduced here, i.e. 71 546. Carrying out the sequence of operations specified above to determine the corresponding binary number gives a table of values (Table 8.1). The result is in the third row, which has to be reversed in order because the most significant bit (the first one) is on the extreme right and it should be on the extreme left. *Note:* decimal fractions can be coped with as well as whole numbers greater than one because they

Operation no.	i	ii	iii	iv	v	vi
Result of division by 2	35 773	17 886	8943	4471	2235	1117
Remaining 0 or 1	0	1	0	1	1	1

Operation no.	vii	viii	ix	x	xi	xii
Result of division by 2	558	279	139	69	34	17
Remaining 0 or 1	1	0	1	1	1	0

Operation no.	xiii	xiv	xv	xvi	xvii
Result of division by 2	8	4	2	1	0
Remaining 0 or 1	1	0	0	0	1

Table 8.1. Conversion of $71\,546_{10}$ (decimal number) into binary: $1000\,1011\,1011\,1101\,0_2$

are represented by negative indices. Thus, e.g. the binary decimal 0.1101 can be expressed as

$$(1 \times 2^{-1}) + (1 \times 2^{-2}) + (0 \times 2^{-3}) + (1 \times 2^{-4}) = 0.5 + 0.25 + 0 + 0.0625$$
$$= 0.8125$$

In Table 8.1, 17 *binary digits* (called *bits*) are needed to specify a number which, in the decimal system, requires only 5 digits. This is not a difficult problem in simulating large numbers by the use of electronic logic gates as is done in a digital computer because of the speed at which such gates can work. If the pulses representing the 'bits' are handled by the logic gates in the central processor of the computer at a rate of, say, one every 10 ns then, within a microsecond (only 10^{-6} s) 100 bits can be processed and in a millisecond (1 ms $= 10^{-3}$ s), 10^5 bits can be processed, which is equivalent to $10^5/17$, i.e. over 5 thousand five-digit (in decimals) numbers. The problem becomes one of obtaining sufficiently fast printers to record the data at the output of the digital computer. Even so, for some digital operations where a printed output is not needed but control (e.g. of an instrument, a plant or a machine tool) is the object of a microprocessor (the central processor of a microcomputer), to be able to speed up the handling of numerically-based data, a hexadecimal system (base or radix = 16) has been developed.

8.4 Alphanumeric Data

The use of logic gates in computer systems is also desirably able to handle the letters of the alphabet (26 in English) as well as the digits 0 to 9 (10). Hence, a system is needed which handles a total of 36 alphanumeric characters. To represent all these bits, there must be available 36 different combinations of 0s and 1s. As $2^5 = 32$, 5 is not enough because 36 characters are involved in such work. So it is essential to go to 6, which gives $2^6 = 64$, and this enables additional symbols to be added to the lower-case (or capital) letters of the alphabet plus the 10 figures. Such symbols can be $+$, $-$, punctuation marks, and on. If it is required to include these symbols and also both capital and lower-case letters, it is necessary to have 7 bits (8 bits are often used) to provide $2^7 = 128$. This means that to reproduce a 'word', at least six bits are needed if a binary system is used and eight is preferable. A 'word' of eight bits is a commonly used unit in digital computer work.

8.5 Truth Tables

Each logic gate (AND, OR, NOT and the other four gates which result from combinations of the basic ones) can have its operation represented by a truth table. So, indeed, can complex combinations of gates. If there are n inputs, there are 2^n different input situations because each input is either 0 or 1 when, as usual, a binary system is employed. In the majority of specifications here (and elsewhere) 3 inputs are considered leading to $2^3 = 8$ possible input situations.

To draw up a truth table with three inputs A, B and C, each now is a possible combination of A, B and C. The values of A, B and C are all either 0 or 1 and there are eight cases. For the output of the gate at the output terminal F, f is used. In the binary system, f is also always either 0 or 1.

8.6 AND Gates

An AND gate can have any number of inputs: $A,B,C...J...N$. Only three inputs are to be considered here. As discussed in section 8.2, the so-called logical 1 and logical 0 are represented by pulses of voltage. The essential feature of an AND gate is that its output f is logical 1, if and only if, *all* its inputs are logical 1. Thus, an AND gate may be considered to be a coincidence circuit: it gives an output only when inputs to it coincide in time. If any of these inputs is logical 0, the output is logical 0. The case of three inputs is represented in Figure 8.2(a).

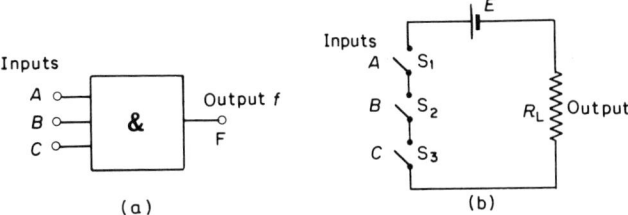

Figure 8.2. An AND gate

An AND operation with three inputs can be represented (as in Figure 8.2(b)) by three switches S_1, S_2 and S_3 in *series* with a source of supply E and a load R_L. To speed things up greatly, the gate is electronic in a single AND gate (Figure 8.2(a) is a conventional symbol). In the AND gate, *all* inputs have to be logical 1 (i.e. all three switches have to be closed if there are three inputs) to give an output voltage. If any one of the inputs is logical 0 (i.e. any one of the switches is open) the output f is 0, i.e. there is no output. For an AND gate, with three inputs, the truth table is Table 8.2.

A	B	C	f
0	0	0	0
0	0	1	0
0	1	0	0
0	1	1	0
1	0	0	0
1	0	1	0
1	1	0	0
1	1	1	1

Table 8.2. Truth table for an AND gate with 3 inputs. *Note:* only when A, B, and C are all 1 is f = 1

Another way of describing the behaviour of a logic gate is by the use of Boolean algebra. In this, the AND operation is represented in the three-input case by

$$f = A.B.C$$

which implies that f is present only if A, B and C are all present. Note that f, A, B and C are each either logical 0 or 1 and that when A, B and C are each logical 1 the result of the 'anding' is not 3 but still only logical 1. Thus Boolean algebra has to be learnt as a new discipline by those familiar only with the classical algebra.

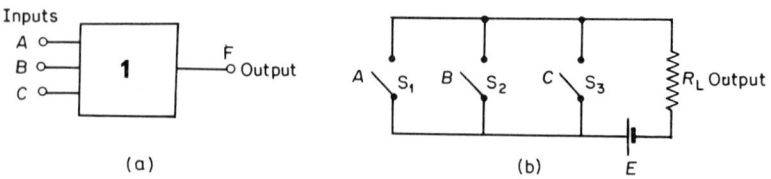

Figure 8.3. An OR gate

8.7 OR Gates

An OR gate with three inputs A, B, C (again, an OR gate can have any number of inputs but only three are considered here) is represented in Figure 8.3(a). As in the case of the AND gate the inputs are each logical 1 or logical 0 and the output f is logical 1 or logical 0. The difference is that now $f =$ logical 1 when any one, or more than one, of the inputs is logical 1. Whereas the AND gate is represented by a simple series circuit of switches, the OR gate is represented by a simple parallel circuit of switches (Figure 8.3(b)). The truth table for the OR gate with three inputs A, B and C (where each of these is logical 1 or logical 0) is given in Table 8.3. The output f is also either logical 1 or

A	B	C	f
0	0	0	0
0	0	1	1
0	1	0	1
0	1	1	1
1	0	0	1
1	0	1	1
1	1	0	1
1	1	1	1

Table 8.3. Truth table for an OR gate with three inputs

logical 0, i.e. either there is an output or there is not. The expression in Boolean algebra for an OR gate is

$$f = A + B + C$$

which implies that f is equal to A or B or C. Again f, A, B and C can each be either logical 0 or logical 1. Note that the . between symbols in Boolean algebra (see section 8.6) implies *and*, not multiplication which is the custom in the usual mathematics and that the $+$ between symbols implies *or*, not addition as is the usual notation.

8.8 NOT Gates: Means of Negation

A NOT gate is also known as an INVERTER gate. It has only one input and one output. The single input may be put as A and the single output as f. Again, both A and f can only each be logical 1 or logical 0. The NOT gate is very simple: its output is an inversion (complement) of its input, if its input A is logical 0, its output f is logical 1; vice versa, if A is 1, f is 0. In other words, f is A negated. The truth table (Table 8.4) is

A	f
0	1
1	0

Table 8.4. Truth table for a NOT gate

so simple as to be hardly worthy of expression. In Boolean algebra, the appropriate expression is

$$f = \bar{A}$$

In this algebra it is the convention that a bar over the symbol implies inversion (or negation) whereby logical 1 is turned into logical 0 and logical 0 into logical 1, as shown in Table 8.4. The symbol for a NOT gate to be used here is as in Figure 8.1(c), the British Standards one.

8.9 NAND and NOR Gates

If a NOT gate (i.e. a means of negation) is inserted at the *output* of an AND gate, a NOT AND or NAND gate is the result. If a NOT gate is inserted at the output of an OR gate, the result is a NOT OR or NOR gate. The symbol for a NAND gate is shown in Figure 8.4(a) whereas for a NOR gate it is Figure 8.4(b). The truth tables for NAND and NOR are given in Table 8.5 and 8.6 respectively. As would be

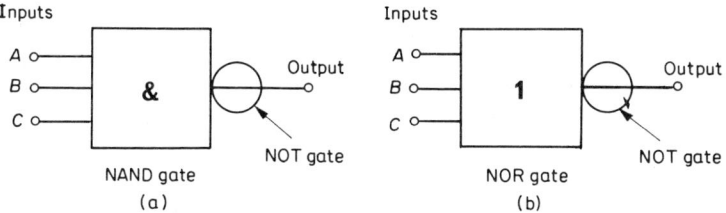

Figure 8.4. A NAND gate and a NOR gate (3 inputs in both cases)

A	B	C	f
0	0	0	1
0	0	1	1
0	1	0	1
0	1	1	1
1	0	0	1
1	0	1	1
1	1	0	1
1	1	1	0

Table 8.5. Truth table for a NAND gate

A	B	C	f
0	0	0	1
0	0	1	0
0	1	0	0
0	1	1	0
1	0	0	0
1	0	1	0
1	1	0	0
1	1	1	0

Table 8.6. Truth table for a NOR gate

expected, these truth tables for a NAND gate and for a NOR gate have outputs f which are the negations of those for an AND gate and an OR gate respectively.

Following section 8.6, the Boolean expression for an AND gate is

$$f = A.B.C$$

so that for a NAND gate it is (see also section 8.8)

$$f = \overline{A.B.C}$$

It is useful to consider a two-input NAND gate of which only one input is used. This is the same as a NOT gate. Indeed, two-input NAND gates, used with only one input (which means essentially that the two inputs are joined so that the same logic level is fed to each), are useful in that they can undertake the same functions as NOT gates. Thus, in integrated circuit form, NOT gates are usually based on two-input NAND gates. They give rise (as shown in section 8.22) to modern digital computer memory elements, based on bistable circuits.

The truth table 8.6 for a NOR gate has outputs f which are the negations of those for an OR gate so the appropriate Boolean algebra expression for a NOR gate with three inputs

$$f = \overline{A + B + C}$$

It can be shown that an AND gate for positive logic is also an OR gate for negative logic.* Thus the NAND (or the NOR) gate can be used as the basis for the whole of the electronics of the central processing unit of a digital computer in all its forms (see section 8.18).

8.10 Exclusive–OR Gates

An OR (or inclusive–OR) gate (section 8.7) has an output f of logical 1 if any one (or more than one) of its inputs is logical 1. An exclusive–OR gate has an output f of logical 1 when one of its inputs is logical 1 but *not* when both of its inputs are logical 1 *simultaneously*.

The exclusive–OR gate usually has only two inputs. Call them A and B. The output f is to be 1 when only one of A and B is 1 but f must be 0 when both inputs are 1 simultaneously. One of several arrangements for doing this (Figure 8.5) consists of two AND gates

Truth table

A	B	f
0	0	0
0	1	1
1	0	1
1	1	0

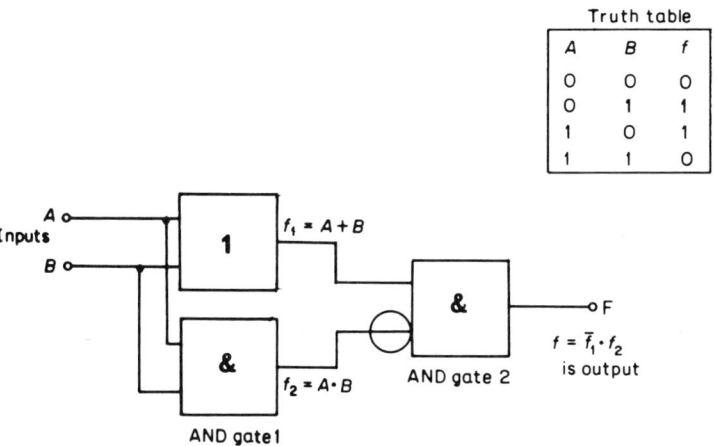

Figure 8.5. A two-input exclusive–OR gate (block diagram of one of several arrangements)

* If we write $V(1)$ and $V(0)$ as the pulse heights of the voltages representing respectively logical 1 and logical 0, in positive logic $V(1)$ exceeds $V(0)$ whereas in negative logic $V(1)$ is less than $V(0)$.

which are called AND gate 1 and AND gate 2, an OR gate and a NOT gate. A and B provide at the output of AND gate 1, $f_2 = A.B$ (to use Boolean algebra) whereas the OR gate (which has the same inputs as the AND gate 1) provides at its output $f_1 = A + B$. The output f_2 is negated to become $\overline{A.B}$ before it is fed to AND gate 2 along with f_1. The output from AND gate 2 is,

$$f = f_1 \cdot f_2 = (A + B) \cdot (\overline{A.B}) = A.\bar{B} + B.\bar{A}$$

Note that the exclusive–OR gate is an inequality detector: it can distinguish between unequal numbers.

Any number (including those less than 1) in decimals can be expressed as a sequence of binary numbers (section 8.3). Two binary numbers will *not* give the same output from an exclusive–OR gate unless the binary values are all the same. Such inequality detectors are useful in the arithmetic and memory-store sections of a digital computer (section 8.18). Exclusive–ORs (i.e. inequality detectors) are available in integrated circuit form. Equality detectors are also used and are available. Both inequality and equality detectors can be built up from appropriate combinations of either NAND or NOR gates. Figure 8.6 shows the conventional symbols for an exclusive OR–gate with two inputs: (a) is the American symbol and (b) is the British standards one.

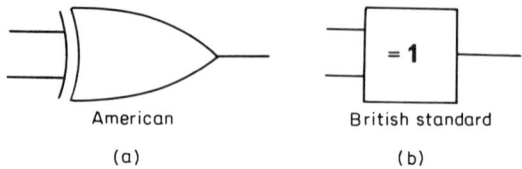

American   British standard

(a)   (b)

Figure 8.6. Symbols for exclusive–OR gates

8.11 Enable Gates

If a NOT gate is inserted at the *input* of a gate, that gate is *enabled*. To illustrate what this means, consider a three-input AND gate (section 8.6) of which one of the inputs, say C, is preceded by a NOT gate (Figure 8.7(a)). The AND gate alone is a coincidence circuit in that all the inputs have to be present simultaneously (i.e. they must coincide) to produce an output. If one of these inputs is preceded — as suggested — by a NOT gate, the input to this NOT gate has to be logical 0 to produce the required logical 1 at the input of the AND gate proper. Hence, the AND gate coincidence action is only *enabled* if there is an input to the terminal of the enabled AND gate which is logical 0. Conversely, if the input to this terminal is logical 1, the

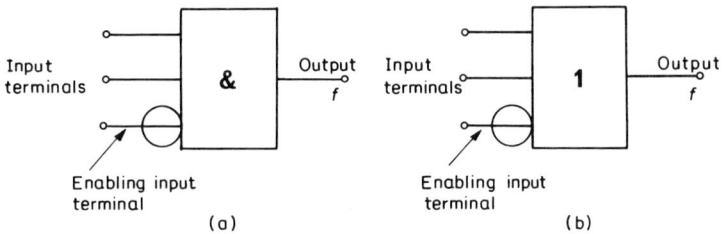

Figure 8.7. ENABLE gates: the use of NOTs at the inputs of (a) an AND gate
and (b) an OR gate

action of the AND gate is inhibited. The truth table for a three-input
ENABLE gate is easily deduced by the reader. An enabling NOT can
also be inserted at the input of an OR gate (as in Figure 8.7(b)).

8.12 Some Experiments to Show that the NAND Gate Provides a Universal Logic Function

Outlined below is a series of simple experiments to demonstrate that
NAND gates alone may be used to perform NOT, AND and OR
functions. It is convenient to use here a 7400 integrated circuit which
contains four separate two-input NAND gates. The pin function of
the 7400 is shown in Figure 8.8(a) but it should be remembered that
the pin numbers are not shown in the sequence in which they occur on
the integrated circuit. The logic gates of the 74 series make use of
transistor-transistor logic (TTL) so that a +5V stabilized supply is
required to power the circuits. Also shown (Figure 8.8(b)) is a simple

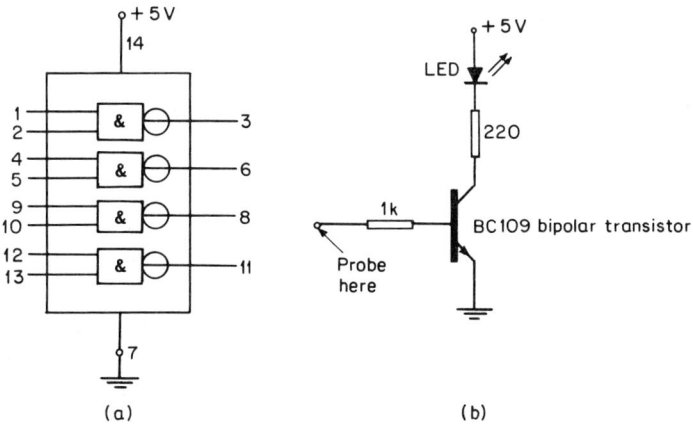

Figure 8.8. (a) the 7400 integrated circuit showing the four 2-input
NAND gates and (b) the circuit diagram of a probe to detect the output state

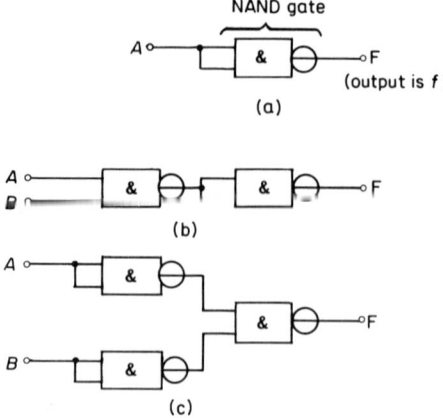

Figure 8.9. Using NAND gates (as an i.c. in Figure 8.8(a)) to provide (a) a NOT function; (b) an AND function, and (c) an OR function

circuit based on a light-emitting diode (LED) in the collector circuit of a BC109 bipolar transistor; this can be used to indicate the output state of the logic circuit in each case.

The NOT function   With the terminal A (Figure 8.9(a)) set first at logical 1 (which is $+5V$) the output is established using the logic probe (of Figure 8.8(b)); with A at logical 0 (i.e. connected to earth), the output is established again. The truth table (Table 8.4) indicates the inverting nature of the gate, i.e.

$$f = \bar{A}$$

The AND function   With the input terminals A and B (Figure 8.9(b)) set at the appropriate logical levels, the truth table for an AND function (Table 8.2, for A and B columns only) is obtained. It is then apparent that inverting a NAND function produces an AND function, i.e. beginning with an AND gate, if a second inversion is carried out, one returns to an AND gate.

The OR function   With the inputs terminals A and B (Figure 8.9(c)) set at the appropriate logical levels, the truth table (Table 8.3 for A and B columns only) can be realized, showing that, in Boolean algebra:

$$f = \overline{\bar{A}.\bar{B}} = A + B$$

8.13 Some Useful Relationships in Boolean Algebra

Conventional algebra is based on the simple laws of arithmetic with the four operations of addition, subtraction multiplication and division. When Boolean algebra is used the operations of Boolean arithmetic must be used and these involve the logical operations AND, OR and NOT. A number of useful relationships in Boolean algebra are listed below. They should be remembered because they enable a rather complex Boolean relationship to be simplified and converted to a form from which a logic circuit can be designed. Any one of the relationships can be verified simply by drawing up a truth table. In any of the relationships shown below A, B and C may be logical 1 or logical 0:

(i) $A + 0 = A$
(ii) $A + 1 = 1$
(iii) $A.0 = 0$
(iv) $A.\bar{A} = 0$
(v) $A + A = A$
(vi) $A + \bar{A} = 1$
(vii) $A.A = A$
(viii) $A.\bar{A} = 0$
(ix) $A.B + A.C = A.(B + C)$
(x) $A + B.C = (A + B).(A + C)$
(xi) $\overline{A.B.C} = \bar{A} + \bar{B} + \bar{C}$
(xii) $\overline{\bar{A}.\bar{B}.\bar{C}} = A + B + C$

The relationships (xi) and (xii) are known as de Morgan's theorem. This is often used to convert an OR type of relationship (either OR or NOR) into an AND type (either AND or NAND). Suppose, for example, it is required to produce an output f where $f = A + B + C$ and only NAND gates are to be used. Using a double inversion (which creates no change)

$$f = \overline{\overline{A + B + C}}$$

Applying (xii)

$$f = \overline{\bar{A}.\bar{B}.\bar{C}}$$

which is easy to achieve in a logic circuit that makes use of NAND gates only and remembering (section 8.9) that a two-input NAND gate serves as an inverter if the two inputs are linked. Some of the Boolean relationships shown above are verified by the truth tables shown in Table 8.7 as (i), (ii), (iii), (iv) and (ix). The reader should draw up truth tables to verify the remaining relationships.

De Morgan's theorem for two variables can be conveniently shown pictorially as in Figure 8.10.

A	0	f
0	0	0
1	0	1

A	1	f
0	1	1
1	1	1

A	0	f
0	0	0
1	0	0

A	$\bar{A}$	f
0	1	0
1	0	0

(i) $A + 0 = A$ (ii) $A + 1 = 1$ (iii) $A \cdot 0 = 0$ (iv) $A \cdot \bar{A} = 0$

A	B	C	$A \cdot B$	$A \cdot C$	$(B + C)$	$A \cdot (B + C)$	$A \cdot B + A \cdot C$
0	0	0	0	0	0	0	0
0	0	1	0	0	1	0	0
0	1	0	0	0	1	0	0
0	1	1	0	0	1	0	0
1	0	0	0	0	0	0	0
1	0	1	0	1	1	1	1
1	1	0	1	0	1	1	1
1	1	1	1	1	1	1	1

(ix) $A \cdot B + A \cdot C = A \cdot (B + C)$

Table 8.7. Truth tables to verify five of the Boolean relationships

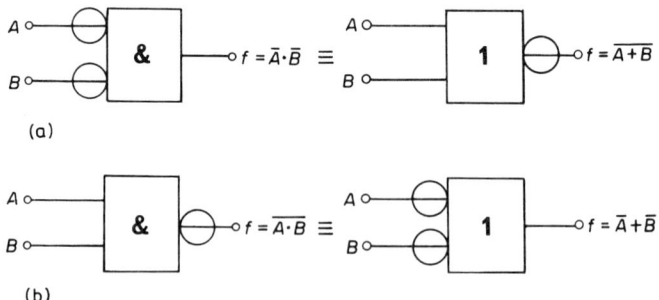

(a)

(b)

Figure 8.10. (a) Showing that the two-gate configurations given are identical is another way of writing $\bar{A}.\bar{B}=\bar{A}+B$ and (b) showing that the two-gate configurations given are identical is another way of writing $\overline{A.B}=\bar{A}+\bar{B}$.

8.14 Designing a Logic Circuit from a Truth Table

Suppose that a truth table has been drawn up to represent the required function. It is possible to proceed systematically from this point to design a logic circuit. The first stage is to derive a Boolean expression from the truth table which will describe the circuit behaviour. Consider the three variables A, B and C shown in Table 8.8. In the last column is written a minterm which is that AND function which includes each variable in the row in true or in inverted form in such a way the minterm can only have the value 1.

A	B	C	Minterm
0	0	0	$\bar{A} \cdot \bar{B} \cdot \bar{C}$
0	0	1	$\bar{A} \cdot \bar{B} \cdot C$
0	1	0	$\bar{A} \cdot B \cdot \bar{C}$
0	1	1	$\bar{A} \cdot B \cdot C$
1	0	0	$A \cdot \bar{B} \cdot \bar{C}$
1	0	1	$A \cdot \bar{B} \cdot C$
1	1	0	$A \cdot B \cdot \bar{C}$
1	1	1	$A \cdot B \cdot C$

Table 8.8.

Now imagine that the truth table given in Table 8.9 has been prepared for a particular logic system and that it is required to design a logic circuit using NAND gates only, so as to produce the output f. The rows marked with an asterisk contain those minterms for which the circuit must produce an output of 1. The output must have a value of 1 if $\bar{A}.\bar{B}.\bar{C}$ or $\bar{A}.B.C$ or $A.\bar{B}.\bar{C}$ or $A.B.C$ is 1. Therefore

$$f = \bar{A}.\bar{B}.\bar{C} + \bar{A}.B.C + A.\bar{B}.\bar{C} + A.B.C$$

This Boolean expression uniquely defines the behaviour. It can be simplified by writing it in the form:

$$f = \bar{B}.\bar{C}.(\bar{A} + A) + B.C.(\bar{A} + A)$$

A	B	C	f	Minterm
0	0	0	1	$\bar{A}\cdot\bar{B}\cdot\bar{C}*$
0	0	1	0	$\bar{A}\cdot\bar{B}\cdot C$
0	1	0	0	$\bar{A}\cdot B\cdot\bar{C}$
0	1	1	1	$\bar{A}\cdot B\cdot C*$
1	0	0	1	$A\cdot\bar{B}\cdot\bar{C}*$
1	0	1	0	$A\cdot\bar{B}\cdot C$
1	1	0	0	$A\cdot B\cdot\bar{C}$
1	1	1	1	$A\cdot B\cdot C*$

Table 8.9.

but

$$(\bar{A}+A)=1$$

so

$$f=\bar{B}.\bar{C}+B.C$$

To convert this expression into one which enables NAND gates to be used a double inversion is performed (leaving the expression unchanged) and applying de Morgan's theorem. Hence

$$f=\overline{\overline{\bar{B}\bar{C}+B.C}}=\overline{\overline{\bar{B}.\bar{C}}.\overline{B.C}}$$

The appropriate NAND gate circuit is given in Figure 8.11.

Example 8.14

A, B *and* C *are three input signals to a control system and* f *is the output. The truth table (Table 8.10 including the minterms), shows the desired operation of the system. Draw a logic circuit using only* NAND *gates to implement this function. (Advanced Level in Electronic Systems, Associated Examining Board, 1976).*

The table is first extended to include the minterms. The asterisks in this table show those rows which must produce a 1. Hence

$$f=\bar{A}.\bar{B}.\bar{C}+A.\bar{B}.\bar{C}+A.B.\bar{C}+A.B.C$$

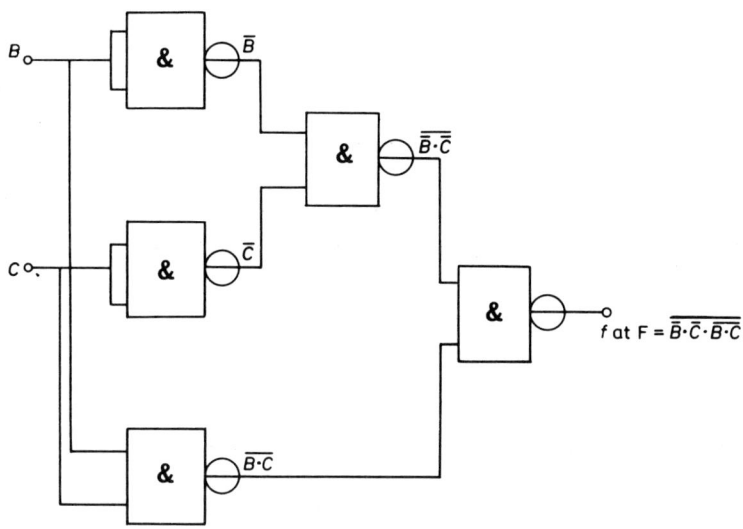

Figure 8.11. Configuration of NAND gates to produce $f = \overline{\overline{B.\overline{C}}.\overline{B.C}}$

A	B	C	f	
0	0	0	1	$\overline{A}\cdot\overline{B}\cdot\overline{C}\,$*
1	0	0	1	$A\cdot\overline{B}\cdot\overline{C}\,$*
0	1	0	0	$\overline{A}\cdot B\cdot\overline{C}$
0	0	1	0	$\overline{A}\cdot\overline{B}\cdot C$
1	1	0	1	$A\cdot B\cdot\overline{C}\,$*
0	1	1	0	$\overline{A}\cdot B\cdot C$
1	0	1	0	$A\cdot\overline{B}\cdot C$
1	1	1	1	$A\cdot B\cdot C\,$*

Table 8.10. For example 8.14 with minterms included

Write

$$f = \bar{B}.\bar{C}.(\bar{A} + A) + A.B.(\bar{C} + C$$
$$= \bar{B}.\bar{C} + A.B$$

Using the double inversion and applying de Morgan's theorem

$$f = \overline{\overline{\bar{B}.\bar{C} + A.B}} = \overline{\bar{B}.\bar{C}.\overline{A.B}}$$

The logic circuit required is shown in Figure 8.12.

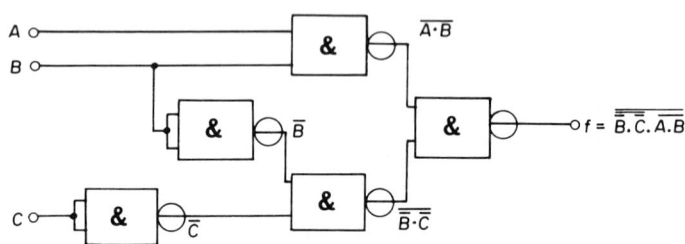

Figure 8.12. The logic circuit using only NAND gates; the solution of Example 8.14

8.15 Pulse Generators; Electronic Switching Circuits

There are three basic types of circuit which are useful in the generation of pulses and to initiate very fast switching action. They are the astable, bistable and monostable circuits, often called multivibrators. Strictly speaking, only the astable circuit is a multivibrator proper: the other two are useful derivations from the multivibrator and of similar circuit design, but they are not really multivibrators. The term 'astable' means 'not stable'. This indicates that oscillations are obtained, although in this case the waveform of the oscillation is non-sinusoidal. Indeed, an output of rectangular waveform is obtained. The astable multivibrator is so-called because it will begin to oscillate (generating an approximately rectangular output waveform) as soon as the supply is switched on. It has two apparently stable (quasi-stable) states and makes a periodic transition from one stable state to the other. The other two 'multivibrators' (the bistable and the monostable) produce pulses which are separated by specific intervals of time when the output is zero: they are generators of rectangular-shaped pulses.

A voltage (or current) of rectangular waveform produced continuously (as by an astable multivibrator) can be shown to be equivalent to the summation of a sinusoidal alternating voltage of a fundamental frequency f plus a very large number of components of frequencies $2f, 3f, ...nf$, where n is an integer. These frequencies are those of the second, third and, in general, the nth harmonic respecively. The term 'free-running' is often used in place of 'astable', when speaking of a 'multivibrator'.

The astable multivibrator

The initial idea (before 1939, with thermionic valves) which gave the astable multivibrator was to use a two-stage resistance-capacitance coupled amplifier (section 4.9) in which the input to the first stage was all the voltage output from the second stage. The feedback would then be 100 per cent (i.e. $\beta = 1$ in equation (4.5)). As each stage of the amplifier produces a change of phase of $180°$ between the input and the output (section 4.14), the overall change of phase with two stages is $360°$. The circuit is consequently an oscillator, but in this case a non-sinusoidal oscillator in that the output is of rectangular waveform.

The circuit of an astable multivibrator based on p-n-p bipolar transistors (p-n-p are denoted here instead of the usual n-p-n, simply to show that either type of transistor can be used provided that attention is paid to the polarity of the supplies) is shown in Figure 8.13(a). The two active components (the bipolar transistors) are not amplifying (the feedback is 100 per cent); they are working as switches with an output which goes negative when the input goes positive and vice versa (Figure 8.13(b) and (c) shows the rectangular output waveforms). The circuit of an astable multivibrator based on an operational amplifier is shown in Figure 8.14(a). The operational amplifier provides an admirably simple circuit in which it acts as the switching device needed to provide the typical rectangular output waveform. Consider that immediately the voltage supply to the opamp is switched on, the output will swing to a saturation value; assume that this is positive saturation at which the output voltage is V_{osat}^{+} (cf. Figure 7.5(a)). A fraction β of this output voltage is fed back to the non-inverting input terminal A. The voltage at A therefore becomes βV_{osat}^{+}, where

$$\beta = R_1/(R_1 + R_2)$$

The capacitor C now begins to charge through the resistor R and when the voltage at X reaches βV_{osat}^{+} (Figure 8.14(c) or some minute voltage above this value) the output of the amplifier will switch to the negative saturation level at V_{osat}^{-} (Figure 8.14(b)). Now the capacitor C will begin to discharge through R until the amplifier output switches again when the voltage at X becomes βV_{osat}^{-}. Assuming that the saturation levels are symmetrical about the common-line voltage, the frequency f of the multivibrator is given by

$$f = 1/[2RC \log_e(1 + 2R_1/R_2)]$$

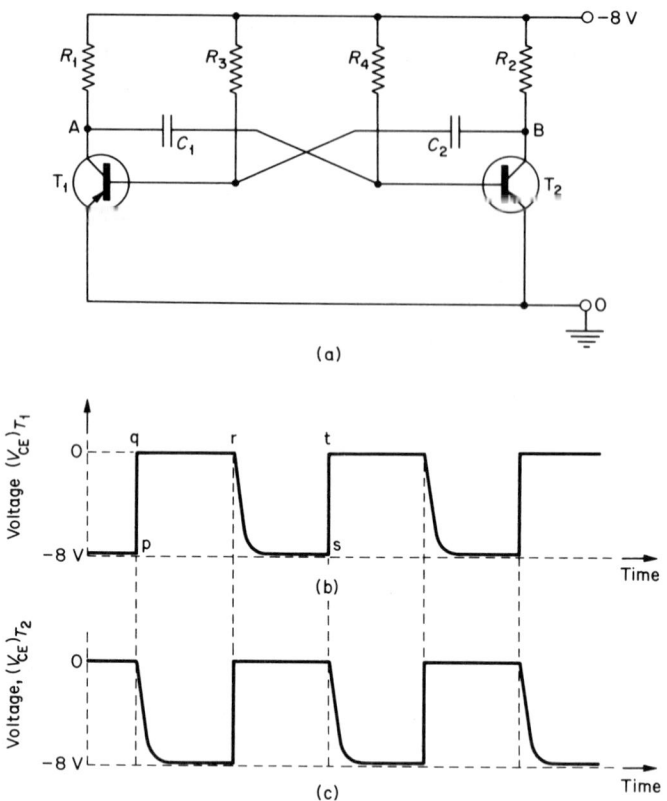

Figure 8.13. A free-running multivibrator utilizing two p-n-p transistors

where f is the reciprocal of the time T of a cycle of the rectangular waveform (Figure 8.14(b)).

Again, as when transistors are used, the opamp is simply acting as a voltage comparator switch and an inverter between two voltage levels. This is what a NOT (i.e. an inverter) gate does (section 8.8); for convenience, NOT gates are commonly two-input NAND gates of which only one input is used. So another way to consider a multivibrator is as two NAND gates (each with only one input) coupled together by means of resistors and capacitors. Integrated circuit forms of all three types of multivibrator (the astable, bistable and monostable) are available and they are recommended for use. Moreover, there is a monolithic integrated circuit waveform generator (type 8038, shown in Figure 8.15(a)) which provides outputs in the form of waves of shapes which are sinusoidal, square and triangular and also ramp (the same as saw-tooth) and pulse outputs of

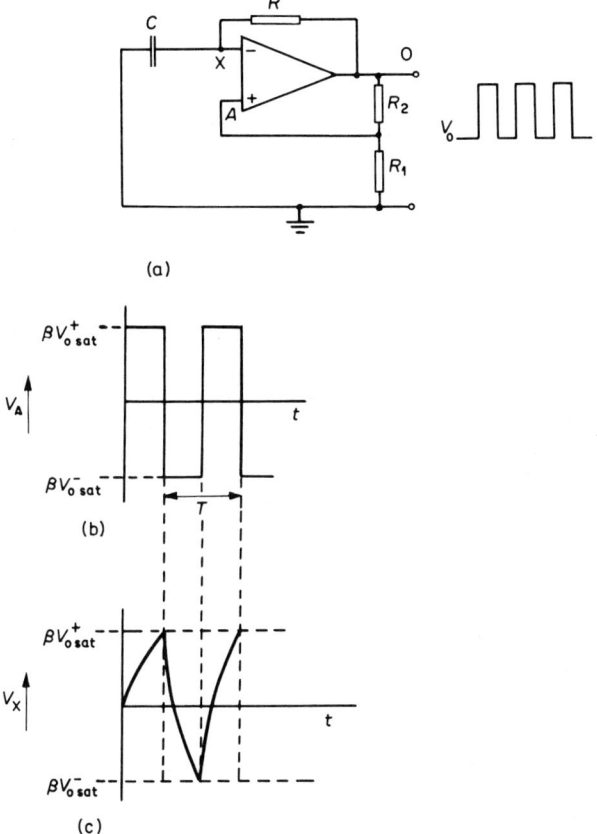

Figure 8.14. An astable multivibrator circuit based on an operational amplifier (Opamp 741; $R_1 = 22$ kΩ; $R_2 = 47$ kΩ; $R = 5$ kΩ; $C = 100$ nF)

which the repetition rate can be set in the frequency range from 0.001 Hz to 100 kHz. Frequency modulation is achievable by a voltage control facility with the 8038. Also, there is a useful electronic timer (type 555 shown in Figure 8.15(b)) in monolithic integrated circuit form which enables an exactly known time to be obtained between two electrically controlled events.

The 8038 waveform generator and the 555 electronic timer each form the basis of a set of useful experiments and simple projects for students who are being introduced to the use of integrated circuits.

The bistable circuit

This is also known as the 'flip-flop' circuit, and is one which can be in either of two stable states and where switching from one state to the

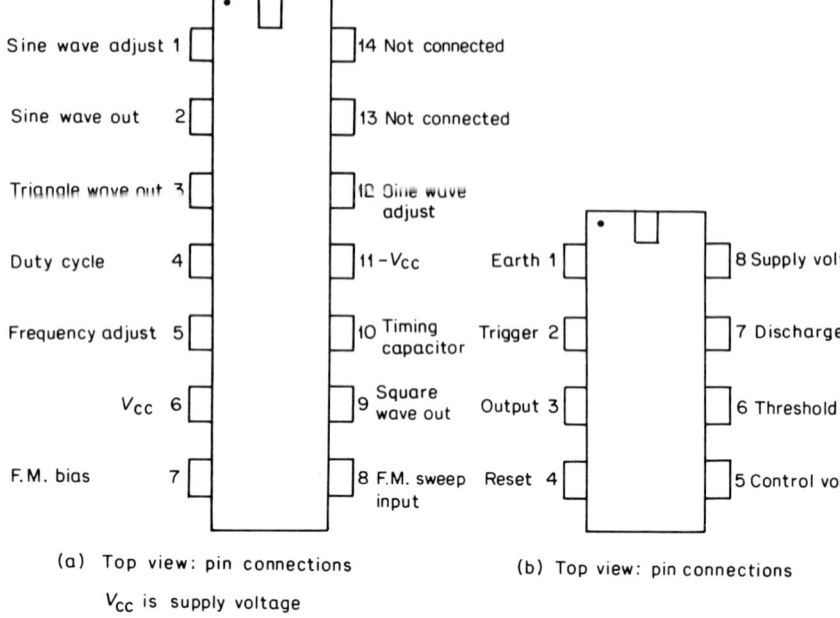

(a) Top view: pin connections

V_{CC} is supply voltage

F.M. = frequency modulation

Figure 8.15. (a) the 8038 waveform generator and (b) the 555 electronic timer

other is by some form of external excitation. By 'stable' here is meant that the circuit can be put into that condition and is able to remain in it for an indefinitely long time, if need be. In electronics, times of operation of 1 μs or even a few ns are needed. An electronic flip-flop circuit is sometimes based on directly cross-coupling two NOT gates: the output from one NOT gate becomes the input to the other NOT gate and vice versa. As NOT gates two-input NAND gates are chosen and, following section 8.9, only one of these inputs is used. The external excitation is in the form of two input pulses, one for each of the states to be taken up by the circuit. Using the alternative, more graphic name for the bistable, the flip-flop, one input pulse causes the circuit to 'flip' into one of the stable states and the next input pulse causes it to 'flop' into the other stable state.

If two pulses are fed into the input of a bistable circuit, the output is a single pulse. The bistable circuit is therefore also a 'scale-of-two': it is a circuit which is able to divide by 2 the number of pulses fed into it. For n bistables following one another in straight sequence, division by 2^n is possible of the number of input pulses to the first bistable, where n is an integer. Four bistables can also be set up in a sequence (not a straight one) which enables division by 10 to be performed, i.e. a decade counting circuit is possible.

The monostable circuit

This is again one in which there are two states and where switching from one state to the other is by an externally applied pulse. Now, however, one of the states is stable and the other is temporary. The monostable circuit remains in the stable state until the externally applied pulse switches it over to the temporary (semi-stable) state. The circuit then reverts (without needing a second input pulse) after a time T_d back to the stable state again. The delay time T_d for the reversion in the case of a monostable electronic circuit is decided by the value of the capacitor and resistor used to couple the second active component back to the first one. This time can be extremely short (1 μs or a few ns) or several seconds. The popular graphic term for a monostable circuit is a 'one-shot' circuit. It provides an output pulse of duration T_d which depends only on R and C and not on the shape or duration of the initiating waveform. The two chief uses of monostables are:

(a) To act as a discriminator in that the output pulse is rectangular and of constant height irrespective of the shape of the input pulse and of its height provided that it exceeds a certain value which will cause the circuit to switch from the stable to the temporary state; hence it discriminates against input pulses of below a specific height.

(b) To introduce a time delay between two pulses. The input trigger is of insignificantly short duration whereas the introduced time delay can be several milliseconds.

8.16 The Schmitt Trigger Circuit

Many years ago (before the Second World War) Schmitt introduced a pulse generating circuit which was triggered by an input pulse. In the Schmitt circuit (originally based on thermionic valves, now always available in semiconductor form) the transition between stable states and the output pulse height is determined by the height of the input signal voltage pulse and is independent of the shape of this pulse. The output pulse height is a constant value provided that the height of the input pulse is between upper and lower values which are close together; in this way it differs from the monostable circuit described in section 8.15. Three applications of Schmitt trigger circuits may be listed.

(i) To maintain constant the heights of the voltage pulses which correspond to logical 1 and logical 0 in the use of logic circuits.

(ii) To provide pulses which trigger into operation other

electronic circuits, which require trigger pulses of specific magnitudes.

(iii) To produce a train of pulses of rectangular shape at the output when the input pulses to the Schmitt circuit have various waveforms.

8.17 The Computer

There are two main classes of computer: the digital and the analogue (see section 8.1). The present concern is only with the much more important digital computer. This may be considered to be in three forms: (*a*) the main-frame computer; (*b*) the minicomputer and (*c*) the microcomputer. This division is, however, simply a matter of convenience: it is really artificial in that the basic principles of all three are essentially the same. The difference between them is largely a matter of size (and application). The main-frame computer is the biggest, it is the batch-processing machine, the 'number-cruncher', much used in data processing especially in commercial work but also in scientific calculations; the minicomputer is medium in size (it is becoming the desk computer for the businessman) and is often an auxiliary to (*a*), especially in scientific and control problems; the microcomputer is the smallest. They are all electronic devices and use at least 1000 times the number of components that a radio receiver does. The microcomputer is based on the microprocessor, which is an extraordinary example of VLSI, based usually on MOSFET (section 4.21). This incredible example of miniature electronics is having a considerable influence on the design and use of the larger computers. Our concern is with the electronics: digital electronics is at the heart of the digital computer. The brief descriptions of this electronics given here begins with an account of the terminology which is common to all classes of digital computer.

8.18 Terminology in Digital Computers

Assembler   Some of the objections to machine programming are overcome by the use of programming in assembly language, in which each instruction is encoded by the use of a mnemonic (definable as an 'aid to the memory' — the human one here) which represents the instruction. A program, called an 'assembler', is stored permanently in the computer's internal memory. Its function is to convert the assembly language into binary form. Such an assembler improves greatly the efficiency with which a program can be written and also makes it easier to spot errors in a program.

Binary digit   Using the first letter of 'binary' and the last two letters of 'digit', this term is abbreviated to *bit*. A group of bits forms the fundamental unit of information which is fed to a computer: it is called a *word*.

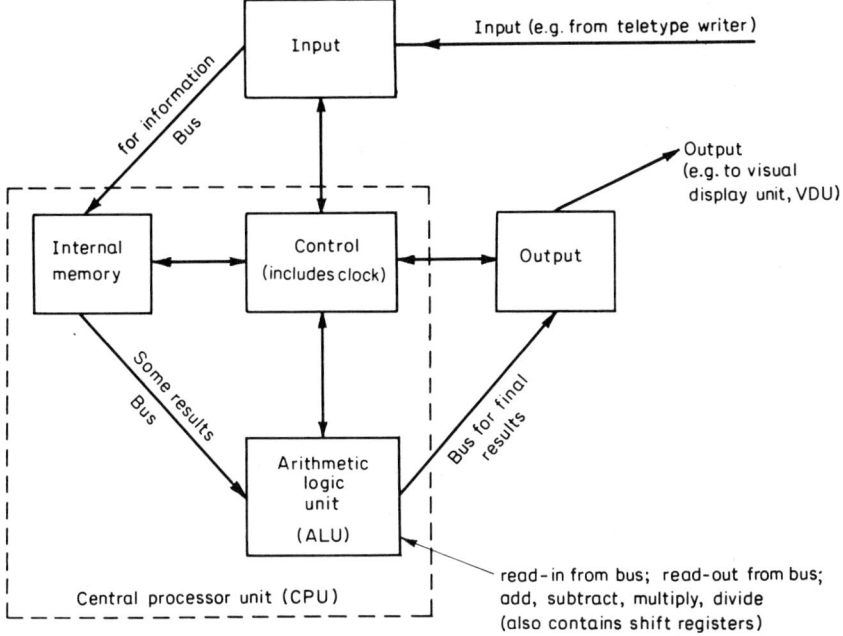

Figure 8.16. Bare outline of arrangement of a digital computer (central processor unit contains ALU, control unit and aspects (sometimes all) of internal memory; peripherals are external store, input and output devices)

Bus This is a single or set of conductors by means of which information in a system is conveyed from one element to another (Figure 8.16). For an 8-bit word there will be 8 parallel conductors in a bus.

Byte A byte consists of 8 bits and can be handled in the digital computer as a single unit. In a 16-bit word there are two bytes.

Central processor unit (*CPU*) This is the part of a digital computer (often in a main-frame machine) which contains the arithmetic unit (based on logic circuitry and usually called the 'arithmetic logic unit': ALU), the control units which involve the internal memory (in which the 'instructions' are stored), and the clock which schedules the way in which the digital computer operates.

Clock A master clock oscillator is a vital part of a digital computer: it is usually based on a quartz-crystal oscillator (see, e.g., section 7.8) of which the exceptionally constant frequency output as a sine-wave is usually turned into a rectangular waveform. The produced train of changes of voltage level (pulses) are thus accurately timed at a specific clock rate or frequency which schedules (time-tables) the operation of the digital computer, usually by making much use of the pulses to operate enable gates (section 8.11).

High-level Languages   Programming in machine language or in assembly language has, the disadvantage that the program which is produced is dependent on the machine it is intended to instruct. A second disadvantage is that these languages do not readily lend themselves to mathematical calculations. To overcome these problems high-level languages have been developed. The structure of each language of this kind is designed to be particularly helpful with the class of problem it is intended to solve. In order to support these languages, a digital computer must have a program — called a 'compiler' and written in assembler — to translate the high-level language into machine code. The high-level programming language can then be used independently of the particular properties of the digital computer concerned.

Typical examples of high-level languages for scientific use are BASIC and FORTRAN. For commercial and business purposes, COBOL is widely used as a high-level language. In attempts to produce high-level languages of greater versatility the high-level languages which have been developed are PL/1 and ALGOL.

Input Data   This is the term given to information received by the computer from the outside world. The devices used for this purpose are mechanical switches, keyboards and teletypewriters. The information is in bit form.

Instructions   A digital computer can only do what it is told to do even though there are *interactive machines* which can be so programmed as to change the sequence of operations if the result of its previous analysis of the data is unsatisfactory. Every single operation together with the sequence in which the operations must be carried out comprises a set of instructions.

Machine language   A digital computer can only work with bits so that any instruction or information given to a computer must be in the form of a code which contains only 1s and 0s. Instructions written in this way (guided by the code book for the digital computer) are said to be in machine language. For example, an instruction may be 10100011 (an 8-bit word). Each 1 or 0 in this instruction has a specific meaning: it tells the digital computer what to do. The computer thus has to be programmed by the use of machine language. However, the process of writing a program in this language is very tedious and it is very easy to make a mistake which it is difficult to trace. Nevertheless, machine programming is efficient and it is still employed with some microcomputers.

Memory   This is also known as *store*, and is a device into which is inserted instructions in the form of bits and which retains or holds these items of information in particular addresses. *The capacity of a memory* is the number of standard words it can retain. The most commonly used memories in digital computers are magnetic-core devices in main-frame machines and semiconductor (solid-state) devices (section 8.22) in minicomputers and microcomputers. There are two main memories in most digital computers: the internal memory (in which are stored the program instructions, including those for addressing and scheduling) and the external memory which is usually capable of holding vast amounts of information (sometimes as much as several million bits) which are not needed as quickly as that from the internal memory. The external memory for a mainframe computer is often in the form of tapes or rigid discs of plastic material coated with a magnetic material which is magnetized and then 'read' by some form of pick-up device; the minicomputer often makes use of magnetic tape, magnetic disc or the so-called floppy disc (a thin plastic disc, similar to the old 45 rpm records, which has a magnetic coating). Microcomputers also use floppy discs, cassettes containing magnetic tapes and semiconductor memory.

Memories, are also classified as *static, random access* and *read only*. A static one contains information stored in fixed locations (addresses) and is available whenever wanted. The internal memory is static: it is based usually on semiconductor bistable circuits (section 8.22) or on magnetic core. A random-access memory (RAM) is such that the 'bit' of information stored in the memory address can be accessed (got at!) by some kind of sensor without any dependence on the location of the previous item of information. One can read from and write into this type of semiconductor memory. This is in contrast to a read-only memory (ROM) which is one where the 'bits' of information are in fixed locations where they are retained permanently. This information cannot be altered in the digital computer by any means. This is useful in that the ROM can store the program which is concerned with the control of a pocket electronic calculator or, in a large computer, is often used for so-called 'microprogramming', which means that certain small parts of a program can be called upon as built into the ROM during its manufacture. ROMs are usually based on MOSFET bistable circuits.

Output data   This is the information sent to the outside world; it is a result of the processing by the digital computer of the input data. Devices used for such output data are the visual display unit (VDU) — which is usually in the form that the output data (in alphanumeric form) is displayed on a cathode-ray tube screen — a teletypewriter (TTY) or a line printer. Note that a teletypewriter (which is an electrical typewriter such that pressing the keys gives out appropriate electrical pulses and vice versa where the reception of appropriate electrical pulses cause the typewriter to type) can operate as either an output or as an input device: it is thus called an I/O device and is connected to the I/O port of the computer.

Program and programmer   A set of instructions which enables the computer to undertake a calculation or data analysis task is called a program. Usually, it is written in a high-level language. The person who analyses the problem is called a *data analyst*. The person (who may also be the data analyst) who prepares the instructions in program form is called a *programmer*.

Sub-routine   This is a short sequence of instructions which is repeated often as part of a program designed to solve a problem by the use of a digital computer. The programmer may well decide to make use of a sub-routine in his program: the sub-routine is obtainable from a program library usually sponsored by the computer manufacturer. The sub-routine forms a particularly useful facility in the programming of data where standard mathematical functions (e.g. that for compound interest calculations) occur frequently. However, it is being replaced by 'microprogramming' techniques based on the use of ROMs.

8.19 Number Systems in Digital Computers; Some Recapitulation

There are three important number systems: the decimal, the binary and the hexadecimal. The first of these is familiar. The second is the binary arithmetic method of representing numerical data and is used in all digital computers because it is so readily simulated by electrical and electronic means based on logic gates (sections 8.2 and 8.3). The binary system may be called the 'base 2 number system'. The hexadecimal system is used on many simple keyboards (rather like those of a typewriter) to input information to a microcomputer. The hexadecimal system may also be called 'the base 16 number system'. This is because the series of numbers which would add up to be equal to the number to be represented (cf. the decimal series in section 8.3) is expressed by powers to the base 16.

As there are only 10 digits in the decimal number system, added means are needed to represent the numbers, 10, 11, 12, 13, 14 and 15. For convenience, the first 6 letters in the alphabet are used, i.e. A, B, C, D, E and F. The strength of the hexadecimal system is the much larger base than in the binary system (16 instead of 2: note, however, that $16 = 2^4$); so many fewer digits are needed to represent the large numbers so frequently encountered in computing. A comparison between the representations of numbers in the three systems is given in Table 8.11.

Example 8.19

(*a*) *Convert the 6-bit binary number 110101 into a number in the decimal system;* (*b*) *convert the 8-bit binary number 10101111 into a number in the hexadecimal system;* (*c*) *convert the 6-bit binary number 111100 into hexadecimal.*

Decimal	0	1	2	3	4	5
Binary	0000	0001	0010	0011	0100	0101
Hexadecimal	0	1	2	3	4	5

Decimal	6	7	8	9	10
Binary	0110	0111	1000	1001	1010
Hexadecimal	6	7	8	9	A

Decimal	11	12	13	14	15
Binary	1011	1100	1101	1110	1111
Hexadecimal	B	C	D	E	F

Table 8.11. Representation of the decimal numbers 0–15 in the three systems

(a) $110101 = 1 \times 2^5 + 1 \times 2^4 + 0 \times 2^3 + 1 \times 2^2 + 0 \times 2^1 + 1 \times 2^0$

$= 32 + 16 + 0 + 4 + 0 + 1 = 53_{10}$

Note that the base 10 — in this case — is conventionally written as a suffix.

(b) Write the binary number in groups of four (because $2^4 = 16$) and convert each group.

$$10101111 = 1010 \quad 1111$$

$$A \qquad F$$

which is written in the form AF_{16} by convention. (The suffix in hexadecimal is clearly 16.)

(c) Arrange, as before, into groups of four from the least significant bit (LSB). (The LSB is the bit on the extreme right-hand side of the binary number — so-called because its contribution to the number is the least.) Add zeros where necessary before the most significant bit (MSB) in the extreme left-hand side to complete the group of four.

$$111100 = 0011 + 1100$$

$$3 \qquad C$$

$$111100 = 3C_{16}$$

8.20 The Binary Half-adder in Digital Computers

As in the case of the addition of decimal numbers, there are a few simple rules for the addition of binary numbers. These are

0 plus 0 = 0

0 plus 1 = 1

1 plus 0 = 1

1 plus 1 = 0 carry 1

The last of these corresponds to $1+1=2$ in decimals which in binary is 10. If A represents one input and B a second, Table 8.2 shows the result of addition. In Boolean algebra, the relationship between the inputs A and B and the sum S can be written

$$S = A.\bar{B} + \bar{A}.B$$

which is the same as the expression given in section 8.10, for an exclusive–OR gate or an inequality detector, of which a diagram is given in Figure 8.5.

Input A	Input B	Sum S	Carry C
0	0	0	0
1	0	1	0
0	1	1	0
1	1	0	1

Table 8.12. The result of the addition of A and B in binary

From Table 8.12, the relationship between the inputs A and B and the carry C can be written, by making use of Boolean algebra, in the form

$$C = A.B$$

which is simply the statement (in Boolean algebra) for the operation of an AND gate (section 8.6).

In the half-adder there is consequently an exclusive–OR gate which produces $(A.\bar{B} + B.\bar{A})$ at S, as shown in section 8.10 and Figure 8.6, whereas an AND gate produces $A.B$ and C, where A and B are the two inputs. Hence, a *half-adder* may be represented by Figure 8.17. Note that the term 'half-adder' is used because the sum and carry are both made available but they have not been combined. This combination is

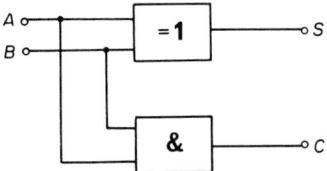

Figure 8.17. A half-adder

given in a full-adder, which is an essential part of the arithmetic logic unit (ALU) of any digital computer. Addition is of great importance in the ALU because multiplication is accomplished by repeated addition, subtraction is achieved by adding the inversions of numbers (obtained by NOT gates) and division is a result of repeated subtractions.

8.21 Removing Contact Bounce in a Mechanical Switch by Means of Two 2-Input NAND Gates

When electrical contact is made (in a digital computer or other system) by using a mechanical switch, the transition is not clean. A train of on–off pulses is produced before the contacts settle. These voltage pulses are said to be due to contact bounce; they can create errors in an electronic system which is fast enough to detect each of these pulses. This is because each pulse can cause spurious switching which occurs a number of times when only one switching action is intended. Such errors due to contact bounce can be removed by the use of cross-coupled NAND gates as shown in Figure 8.18. For a two-input NAND gate, the truth table is Table 8.13 (see also Table 8.5

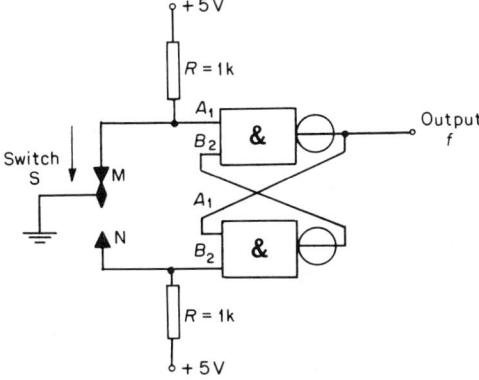

Figure 8.18. Removing contact bounce from a mechanical switch S by means of two 2-input NAND gates

A	B	$f = \overline{A \cdot B}$
0	0	1
0	1	1
1	0	1
1	1	0

Table 8.13. Truth table for a two-input NAND gate

which is for 3 inputs). This Table 8.13 shows (as would be expected) that the output is logical 0 only if both the inputs are logical 1; if one input is at logical 0, the output is at logical 1, regardless of the state of the other input.

In the circuit of Figure 8.18 for removing the contact bounce from the switch S, when S is in position M, the output from the circuit is f = logical 1; for reliability, any such unused terminal to a logic gate should be connected to a source of logical 1 (here $+5V$). This is done in the circuit by means of the resistor R which connects M to the $+5V$ terminal; R has a value of 1 kΩ. When the switch S is depressed the earthed point on it moves away from the contact M and terminal A_1 rises to logical 1 but the output remains unchanged. As soon as the switch S reaches contact N, the terminal B_2 becomes logical 0 and the output f switches to logical 0. Having achieved this switching action, the output f cannot alter even if B_2 does not remain at logical 0 because of contact bounce. This fast transition from logical 1 to logical 0 is ideal for digital systems; a circuit of this type is always used between a mechanical switch and a digital system.

8.22 The Set–Reset (S–R) Bistable Circuit using NAND Gates

In the context of the digital system in general and the digital computer in particular, the word 'bistable' (section 8.15) means a device which can remember or store in its memory a bit, whether it is in logical 0 or 1 state. This device is a particularly important element in a semiconductor static memory of a digital computer. Static memories retain the stored information indefinitely, provided that the power supply is maintained. When the computer is switched off, the information in the store is lost. In microcomputers, particularly those used by the hobbyist, the programs are permanently stored in the external memory consisting of the magnetic tape of a cassette. They are then copied in a few minutes into the internal memory of the semiconductor bistable kind when the computer is switched on.

A set–reset (S–R) bistable circuit for a single stage of such a semiconductor bistable memory makes use of two 2-input NAND

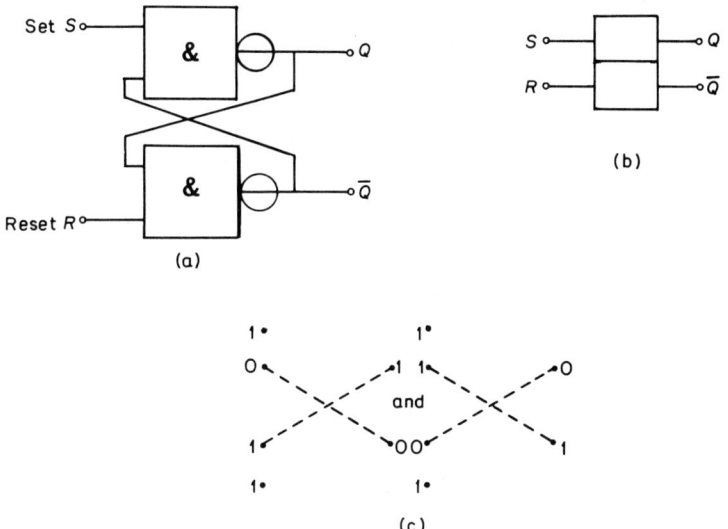

Figure 8.19 (a) An S–R bistable based on two 2-input NAND gates, (b) the circuit symbol for an S–R bistable, and (c) illustrating two possible stable states with both S and R set at logical 1

gates: it is shown in Figure 8.19(a) where the general circuit symbol for an S–R bistable (flip-flop) is shown in Figure 8.19(b).

The output from the bistable is labelled Q in the diagram of Figure 8.19(a). A second output is also available: it is always the complement (negation) of Q, i.e. $\bar{Q}$. With both the set and reset inputs at logical 1 level, the circuit is in a stable state but this state is not known because both the patterns indicated in Figure 8.19(c) are equally valid. However, a negative pulse (a sudden transition from $+5V$ to $0V$) applied to the S input sets Q to the logical 1 state whereas a negative pulse to the reset input R resets Q to the 0 state. Trying to set or reset the bistable to the state it is in already, will obviously not affect the output. Once switched to the desired state, the circuit will remain in that state with both the inputs at the logical 1 level. Hence, this is an example of sequential logic in that the state of the bistable depends on what happened to it previously and not only on the present state of the inputs.

It should be noted that with the bistable circuit shown in Figure 8.19(a) both inputs S and R must never be allowed to be simultaneously in a logical 0 state, for this would imply that both the outputs would be at logical 1 (giving $Q=\bar{Q}=1$ which is logically impossible).

8.23 A Clocked S–R Bistable Circuit

The circuit of a clocked or gated S–R bistable is shown in Figure 8.20. The clock and the set S form the two inputs to a NAND gate *A* whereas the clock and reset R form the two inputs to NAND gate B. As a result of the inversion (with two inputs at the same logical level, the output Q is the inverse) of these two NAND gates, the set and reset

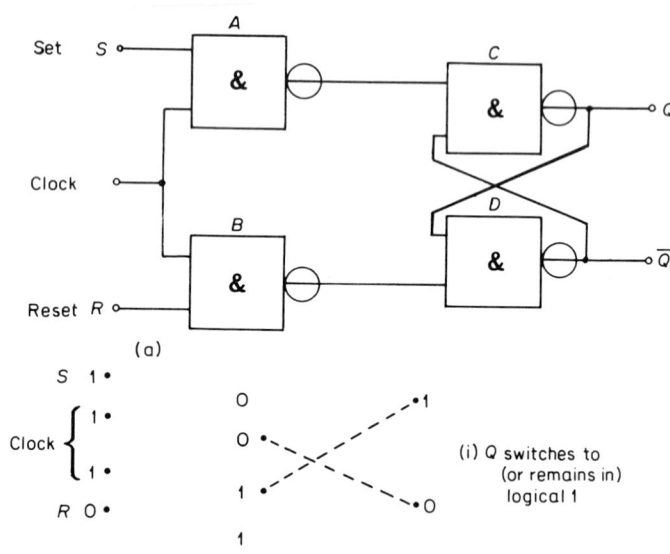

(a)

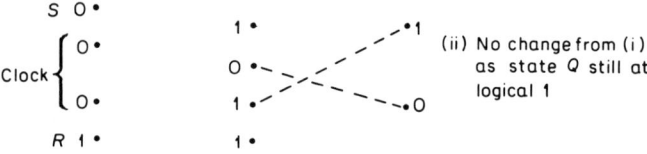

Truth table for clocked S–R bistable with S and clock simultaneously at logical 1 and R at logical 0

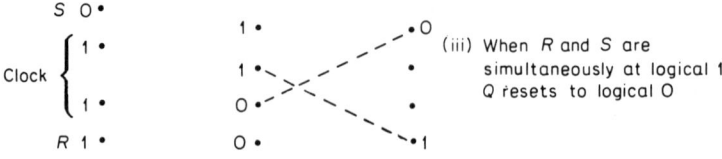

With S at logical 0; clock at logical 0; R at logical 1

With S at logical 0; clock at logical 1; R at logical 1

(b)

Figure 8.20. (a) A clocked or gated S–R bistable and (b) truth tables for states (i), (ii) and (iii), also showing switching that occurs

inputs must now be normally at logical 0. The clock input is also normally at logical 0. To set the bistable to the $Q = 1$ state, the set input is switched to logical 1 but nothing happens until such time as the clock input also goes to logical 1. When this occurs the output Q from gate A goes low and the output Q switches to (or remains in) the $Q = 1$ state. Thus, the S and R inputs determine what switching action should occur but the clock pulse controls the time at which the switching action takes place. This clock pulse is derived from a quartz-crystal oscillator (section 7.8) because it provides such accurate timing* and so determines the sequence of events or can be said to orchestrate them. The use of clock pulses to time the sequence of events is of the utmost importance in all forms of the digital computer.

8.24 Microprocessors and Microcomputers

The fuller account here is of the smallest digital computer. This is because most of the recent advances in the electronics (and use) of the larger computers have resulted from developments in the microcomputer. The microcomputer is based on the microprocessor, which is the most important single product of the microelectronics revolution: it is a single highly complex, integrated circuit which is effectively the heart (the CPU) of a very small computer. As it is the product of very large-scale integration (VLSI), this 40-pin integrated circuit of a microprocessor contains several thousand components, is extremely reliable, demands little power for its operation and is cheap enough to be built into toys and games.

The microprocessor became a reality as a result of the combination of the technological development of VLSI with the concept of the stored program. Unlike the electronic calculator (the so-called 'pocket calculator'), however, the microcomputer (based on the microprocessor) can be programmed to undertake problems of various kinds whereas the electronic calculator makes use of pre-set ROMs to the extent that it is already programmed to undertake specific tasks (e.g. trigonometrical functions, logarithms, square-roots, error functions, etc.) which cannot be altered at the will of the programmer. In early digital computers (including the transistorized ones, and all of the main-frame type) the simple operation of adding two numbers required an elaborate circuit containing many logic gates and a program (instructions telling the computer what to do) which was written by the programmer and based on a language (code) such as FORTRAN. This program was usually fed into the internal memory of the computer which was often a magnetic-core device. In the microprocessor, these logic gates have been largely replaced by a stored program, some of which is held in the memory which is based on a semiconductor circuit (cf. section 8.23) and is on the microprocessor chip. Because a memory is capable of storing several thousand words (where each 'word' is a string of 'bits'; the length of the word — often 8 or 16 bits in a microcomputer) and each word can replace a logic gate, the number of integrated circuits (the so-called 'chip count') for a microcomputer falls dramatically. As a result, the manufacturing costs fell and the reliability of the microcomputer system was improved. A clear distinction needs to be made between the microprocessor and the microcomputer. The microprocessor is a complex logic integrated circuit unit containing memories, counters, decoders and an arithmetic logic unit (ALU). The microcomputer makes use of a microprocessor as a central processing unit (CPU — all digital computers have a CPU) but also contains additional memory chips which store information, interface adapters to link the system with the outside world via an input/output (I/O) port and a master clock to schedule the operations.

Exercise 8

1. Write a short account — with examples — of the use and importance of the terms 'analogue' and 'digital' as used in electronics. Give one important example of the advantage of digital methods over the analogue ones.

2. What are 'logic gates'? Describe in words, the appropriate Boolean algebra and with truth tables (with three inputs) the principles of the following gates AND, OR and NOT. Do *not* consider details of the circuits involved, but draw the conventional symbols for the gates.

3. A digital computer is able to deal with 5×10^7 bits per second. How long will it take for such a computer (assuming there are no interruptions) to undertake the electronics operations (not including the operation of the input and output devices) needed to deal with 1000 words of information, assuming that each word is of average length equal to 8 bits?

4. (a) Determine the binary numbers which correspond to the decimal numbers (i) 86 003 and (ii) 274 019.
   (b) How are decimal fractions considered in binary form?
   (c) Determine the decimal numbers which correspond to the binary numbers (i) 100110 and (ii) 10011110.

5. Using binary digits explain concisely how it is possible to handle information which is alphabetical. Why is a 'word' (in computer language) often chosen to be eight bits?

6. Explain the terms 'NAND gate' and 'NOR gate'. Also explain with appropriate symbolic diagrams how the operation of any of the three basic logic gates: AND, OR and NOT, can be produced by the use of

NAND gates alone. How is an ENABLE gate arranged? What are the chief functions of ENABLE gates?

7. Explain what is meant in relation to pulse generators, by the terms: *astable*, *bistable* and *monostable*. What is the chief function of the last of these three and in what important manner does it differ from the Schmitt trigger circuit in this respect? (Circuit diagrams of pulse generating circuits are *not* required.)

8. In relation to the *digital computer* give definitions of the following terms: machine language; assembly language; bus; central processor unit; clock; high-level programming languages; memory; ROM and RAM.

9. Write an account, with an appropriate diagram, of a clocked set–reset bistable circuit as an element in a memory circuit able to store bits.

10. Write a short essay about the microprocessor and the microcomputer.

Appendix

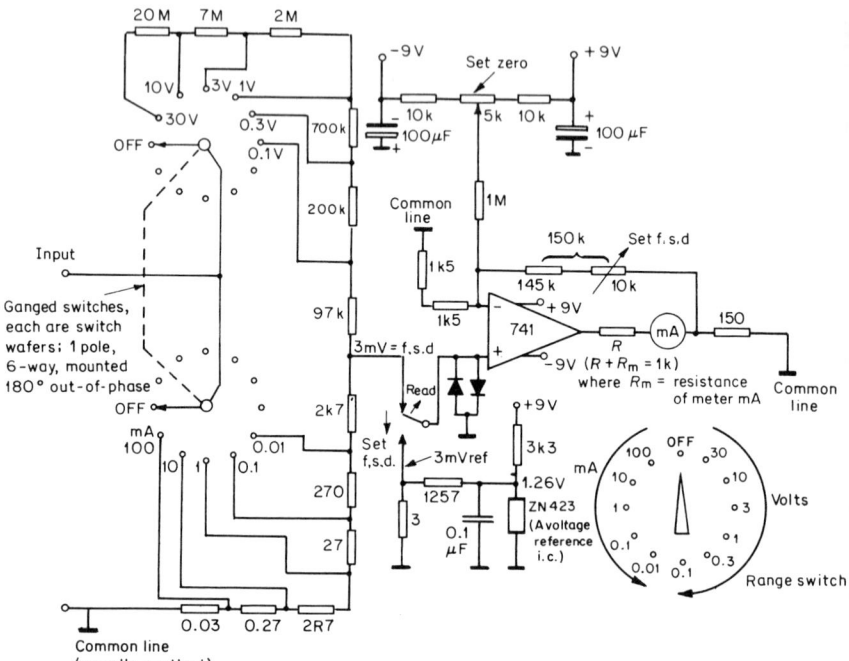

Circuit diagram of a multimeter which can be built as a students' project.

Description of Figure: A multimeter based on an opamp of type 741, battery-operated $(-9, 0, +9$ V). The centre of the battery at 0 V is connected to the common line. On–off switch for the battery is not shown. Pins 4 and 7 of the opamp 741 are connections for the $+$ and $-$ of the power supply voltage — in this case $+9$V and -9V respectively. To obtain small resistance values of 0.03 and 0.27 Ω shown in the bottom left-hand corner of the circuit diagram, use may be made of the copper strip of Veroboard which has (via the copper) a resistance of approx. 1 milliohm (0.001 Ω) between neighbouring holes. As a voltmeter this instrument has the very high resistance of 1 MΩ/volt. (Note that, for example: 7M $= 7\,000\,000$ Ω; 10 k $= 10\,000$ Ω; 2 k7 $= 2.7$kΩ; 27 $= 27$ Ω.)

208

Answers

Exercise 1

3. 3.5×10^7 m s^{-1}
4. 0.454 kms^{-1}; 8.5×10^{-5}
5. 7.5×10^{-2} m
10. 0.282 nm
15. 3.8×10^{20} m^{-3}
16. 2.5×10^{19} m^{-3}
17. 3.2×10^{21} m^{-3}

Exercise 3

6. 141 m A
7. 1:0.81
8. 30 μF

Exercise 4

12. 2.0 V; 2.0 mW
15. 0.0018 μF

Exercise 5

2. 1.65×10^{-6} m

Exercise 6

2. 3.98 pF; 39.8 μs

Exercise 8

3. 1.6 ms
4. (a) (i) 0100111111110011; (ii) 000010111001100011
   . (c) (i) 38; (ii) 158

Index